Raising the Homestead Hog

Raising the Homestead Hog

Jerome D. Belanger

Rodale Press
Emmaus PA

2 4 6 8 10 9 7 5 3 1

Printed on recycled paper

Library of Congress Cataloging in Publication Data

Belanger, Jerome D.
Raising the homestead hog.

Includes index.
1. Swine. I. Title.
SF395.B35 636.4 76-55010
ISBN 0-87857-132-9

CONTENTS

INTRODUCTION

For a genuine, homestead-produced bacon-lettuce-tomato sandwich, first you must get a pig.

Some people are skeptical about the idea of raising a pig in the backyard. After all, they point out, pigs are smelly and dirty. And if real hog farmers with all their automated feeders and automated manure-handling facilities and fancy housing and feeds and antibiotics and volume buying and selling still have trouble making money, why should we presume that a one-hog operation would be profitable? And finally, don't you just about need a college degree in pig butchery to do the slaughtering, cutting, curing, and smoking?

There might be a grain of truth in these notions, but they aren't necessarily 100 percent true. Swine undoubtedly fit into far more homestead and small-farm situations than most people realize. There are some mighty good reasons for raising your own pork, especially if you're interested in the organic way of living.

What's involved?

We got our first pig when we lived on a one-acre homestead right on the edge of a small town. There were no restrictions on livestock in this small agricultural community, and we also raised goats, sheep, geese, rabbits, chickens, and pigeons. Although it's hard to put a value on each individual enterprise in a situation like this, I always had the feeling that hogs were one of the more profitable projects.

Did the neighbors complain? Not at all. In fact, one right across the road kept a couple of sows in his garage. We bought feeder pigs from him, and keeping the newly weaned pigs in our pen became one of our biggest problems—they kept getting out and running across the road to mama!

When we began farming in earnest, we chose hogs. Swine are ideal for capital-deficient, part-time farmers like me, for reasons we'll examine in more detail later. In a sense, they require less experience than most other sectors of farming. It's easier to get set up in hog farming than in, say, dairy or beef farming. While it might be incidental, if you like pigs as much as I do, you'll find they're fun to work with—friendly, intelligent, and full of personality. And, although the hog business is notorious for cycles, hogs can be profitable.

With what I've learned both as a one-hog homesteader and as a commercial hog farmer, I'm convinced that hogs can fill an important place on the organic, labor-intensive homestead—whether to produce meat for home consumption, just a little extra to sell to friends and neighbors, or as a profitable part-time job for a person with acreage and a yen to farm as a sideline.

For the family with the proper location and resources (and this doesn't necessarily mean acres of land, as witnessed by my first setup and my neighbor with sows in his garage), home-produced pork can not only be a very economical source of animal protein: it can be just about the most delectable meat you'll ever set on your table. While these two factors alone are enough to interest many homesteaders in hog raising, there are others.

For example, hogs can reduce waste, which is extremely important on the well-managed homestead. Every gardener has trimmings, thinnings, produce that isn't quite up to kitchen or market standards, and just plain surplus. These items can go on the compost heap of course, but why not feed them to a hog, and then put the hog manure on the compost heap? Hogs also make excellent use of other homestead surplus, particularly eggs and milk (or skim or whey). In addition, pork is a prime ingredient in all kinds of sausage, which means the availability of pork will help the homesteader make better use of scraps and less desirable portions of meat butchered from other stock, including cull milking goats and cull breeding stock from the rabbitry.

Butchering a pig really isn't all that difficult. Certainly for the homesteader who has worked his way up through chickens, rabbits, and perhaps kid goats, slaughtering and cutting up a hog isn't entirely new. Although I can't prove it, I strongly suspect home butchering has something to do with the exquisite quality of homestead pork. This pork is so unlike the store-bought variety that once you taste it, you'll compare it with homegrown vine-ripened tomatoes versus the watery, green-picked ones shipped north in the winter.

There's lots of room for creativity in home curing, smoking, and sausage making, and here again the homestead product can't compare with any but the most expensive commercial ones. Yes, these tasks require a certain amount of knowledge and experience, but they're nothing the average person can't master. Once you get the hang of it you'll not only enjoy meat products that can't be bought today at any price, but you'll also have the satisfaction of knowing you produced it yourself, plus the very real security of *knowing* you know how, should the need or desire arise again.

There are infinite variations in sausages, there are numerous curing methods, there are countless ways to cut the pork carcass. You'll probably make a few mistakes the first time. But consider that in former times young people used methods and recipes handed down through generations, and they went through an apprenticeship beginning at a very early age. Should your training be in mechanics or accounting or nuclear physics, don't feel bad if you don't enjoy gustatorial success the first time you try to cure a ham; in spite of your educational and cultural advantages, you're simply underprivileged when it comes to knowing how to produce good pork products. This book and a little experience will help correct that.

Hogs on the homestead

For the homesteader interested in an agricultural sideline—an income-producing project that can be worked into the busy schedule of homesteading plus an outside job—hogs are probably an ideal project. There is a ready market in nearly all parts of the country, which is not the case with most other common homestead products. Many homesteaders would love to raise rabbits for meat on a commercial scale, or to sell goat milk, but they usually discover that there simply is little market for such products. Or, homesteaders might have garden produce to sell—when everybody else has garden produce to sell, too, so that the stuff is just about worthless.

Also, there is no great start-up expense, as there would be for any kind of commercial dairy, for example. And hogs do not require the meticulous attention that dairy animals do. Moreover, you can be exceedingly flexible even after having decided to raise hogs: you can have a few, or many; a feeding operation or a farrowing operation; you can be labor intensive or capital intensive; you can raise hogs just during the summer and spend the winter by the fire. You could even decide to raise hogs this year and not next year, without taking the financial pounding you'd probably face in most other enterprises.

In fact, hogs have traditionally been raised in small numbers as a part-time enterprise. Although huge swine complexes have arisen in recent years—and they do have *some* affect on the smaller operations—there are still many opportunities in hogs for the small family farmer. In fact, there are some advantages to being small. The legendary fluctuation in the hog market can work to the benefit of the small, flexible producer who is capable and knowledgeable; also, the smaller or single owner/operator stands a better chance of making a profit in a bear market than the fellow with a lot of overhead.

And finally, hogs are the most prodigious producers of one of the best barnyard manures. This presents a financial problem to large confinement operations. But if the homesteader or small farmer isn't downright thrilled about the bounty of black gold that has to be shoveled, he should at least be appreciative of the free fertilizer.

Pigs, in other words, have something for nearly everyone. Among other things, that means this book covers certain facets of swine production that can't be expected to interest or appeal to everyone. For example, if you have no intentions of becoming a commercial hog producer, you'll spend less time with commercially oriented sections than you will with those dealing with homestead hog raising. However, I'd like to point out that any small producer can usually pick up ideas from larger operations, even if management methods do differ. With this in mind, this book is organized into three sections. The first chapters provide general hog information that will give a background for more detailed discussions. Chapters 4 and 5 deal with specific approaches to hog raising, from the one-hog homestead to complex commercial operations. And chapters 6 through 12 deal with details that can be applied to any type of hog enterprise: feeding, diseases, and other management topics. The Appendix covers farrowing, castrating, the legalities of raising hogs, and a brief discussion of the nitrate/nitrite issue.

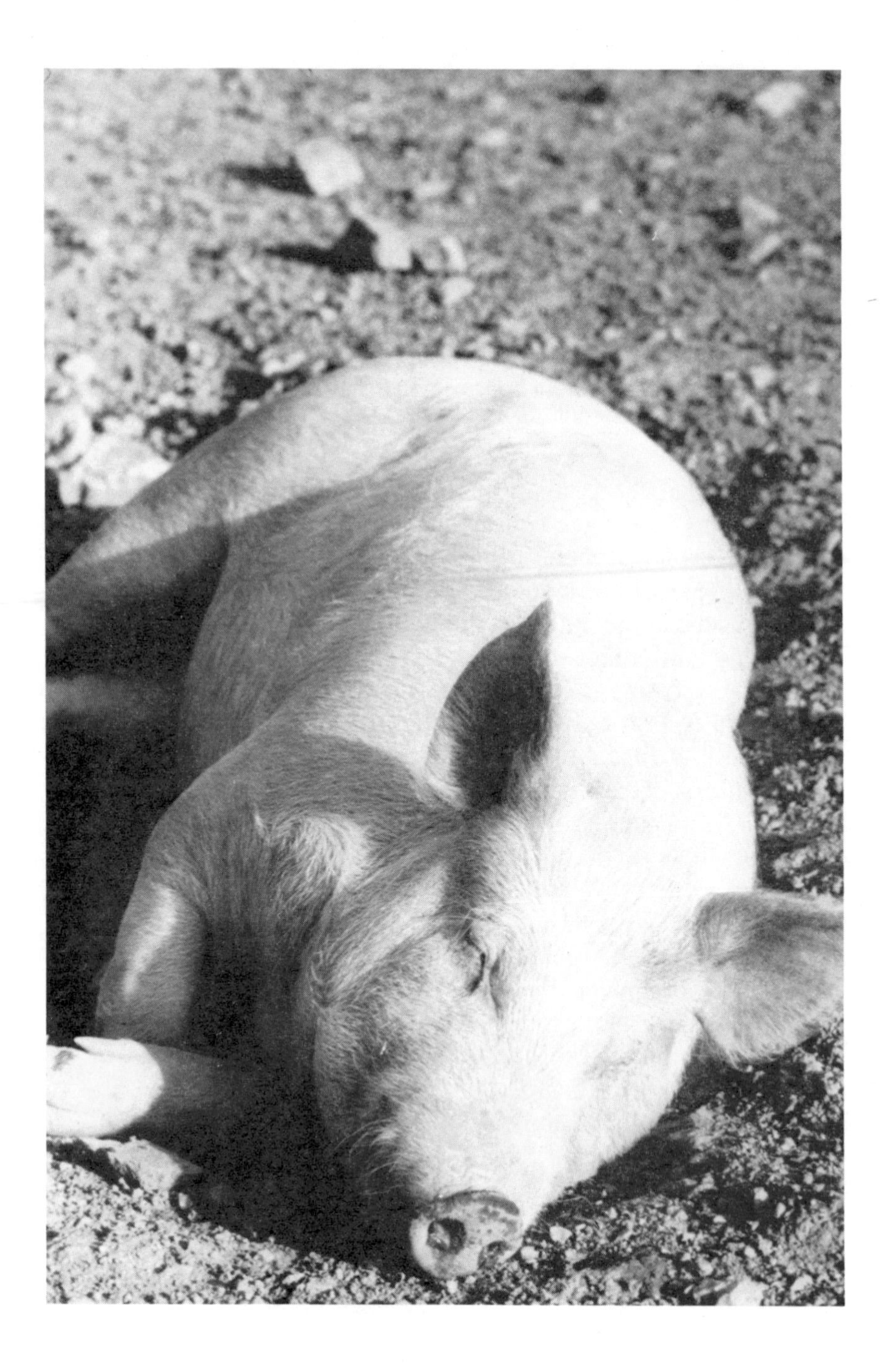

Chapter 1.
THE BASICS

Getting started

When city-bred people decide to go back to the land, there's a lot of learning to catch up on. They quickly learn that the simple life really isn't very simple. Agrarian peasants might not have had much schooling, but they were certainly educated—as we learn in a hurry when we try to be just a *little* self-sufficient!

Because of this, most modern homesteaders take things slow and easy, learning as they go and trying not to make more than one mistake at a time. They also do a great deal of study and research. In addition, many of them consciously or unconsciously follow a certain pattern that closely parallels the development of agriculture itself. Homesteading frequently starts with an interest in natural foods, and then gardening. Chickens and rabbits add to the self-sufficiency, and a dairy animal (cow or goat) comes later. Only then, in the normal pattern of events, do homesteaders seriously begin to think in terms of "real" meat animals, real meat in our society being beef, pork, and to some extent, lamb.

This is a logical progression from several standpoints. It places the greatest emphasis on the most important food groups, but of more importance, it takes the city-wise but country-innocent person through a step-by-step educational process. Gardening requires work and knowledge and skill gained only through experience. In addition to developing skills, gardening shapes people's attitudes and offers some very distinct rewards. All of these can be an enticement, and a help, to small-stock raising. Tending small stock, in turn, further develops the budding homesteader by bringing in an awareness of animal health, genetics, nutrition, housing, and much more, as well as generating increased satisfaction and the sense of security that comes from increased self-sufficiency and knowledge.

As the homesteader becomes familiar with the demands and the rewards of working with living animals, he or she can feel more confident about working with larger stock, which is more expensive, requires more capital outlay for housing and equipment and feed, takes more space, takes a longer time to "harvest," and is harder to handle.

Not everyone falls into this pattern, of course, but it *is* common, and the person with such a background will discover that he already knows more about raising pigs than he thought he did. At the same time, a person with that much experience is likely to take a pretty realistic view of a new venture such as hog raising.

The best course comes from being aware of all the possibilities, by being creative and open-minded enough to search out and recognize new ideas, and by being smart enough to come up with a blend just right for the individual situation. This means that the small-scale hog producer needs to know just as much as the swine tycoon.

When you know what you're doing and you do the job right, you get not only better quality and more efficient production, but more satisfaction. This cannot be overemphasized.

No one does well at something he doesn't enjoy. That's one reason you find people raising hogs in the middle of dairy cattle country, or milking cows in traditional beef territory. They *want* to raise hogs, or dairy cows, even though local markets and conditions might favor another enterprise. Even the choice of breed enters into this: one farmer prefers one breed, his neighbor prefers another, and usually for very personal, intuitive reasons. They simply *like* their breed for whatever reasons they may give, and because they like it, they do better with it.

If this matter of choice is important to the full-time farmer, how much more important is it to the person who works hard all day at one job and then comes home to face another? We don't hear much about this in the context of scientific, profit-oriented management, but the personal factor will be important as long as people remain human beings. In other words, don't expect to be very successful with pigs unless you sort of like the critters.

With all of this in mind, let's go back to the very beginning and get familiar with this animal.

A brief history of the hog

For starters, hogs are members of the kingdom Animalia; the class Mammalia (they are warm-blooded, hairy animals that produce their young alive and suckle them on a secretion from the mammary glands); the order Artiodactyla (even-toed, hoofed mammals); the family Suidae (wild and domestic swine but excluding peccaries); the genus *Sus* (now restricted to the European wild boar and its allies and the domestic breeds derived from them); and the species *Sus scrofa* and *Sus vittatus* (the European and East Indian hogs that contributed to modern domestic swine).

Hogs were known in ancient times. Early nomadic herdsmen didn't make much use of them because they weren't as easy to move as cattle, sheep, and goats. Early swineherds were held in contempt, probably due to the aroma of their stock. However, swine became an important class of livestock early in China's history, and some people believe the Chinese continual regard for hogs, and hog manure, played a large role in that country's agricultural success over all those generations. There is evidence that swine were domesticated before 6000 B.C. in Asia. They have been the most important class of livestock in China for ages, and pork is traditional in the Chinese diet.

There is much more recent evidence of the importance of swine to agriculture. In the early 1970s, one of the goals of South Vietnam's leaders was to produce five tons of rice and two hogs with one worker per year per hectare (2.471 acres). The hogs were important in this plan, not so much for the meat as for the fertilizer that would make the five-ton yield of rice possible. (A 110-pound pig produces a ton of manure per year.) There has to be a lesson there. Not only is the fertilizer produced from renewable resources instead of petroleum (as many chemical fertilizers are), but also there is no transportation expense, the fertilizer factory reproduces itself, and after you get the fertilizer you can eat the factory. How can you beat that for a complete natural cycle?

Archaeologists tell us that the European wild boar was domesticated in northern Europe during neolithic times. Some of the animals still roam the forests of Europe. This was the pig renowned throughout history as the quarry in the elitist sport of boar hunting. His strength, strong tusks, great ferocity, and fighting ability made him a prize trophy for those hunters who dared to test their powers against such a beast. By custom, the boar was hunted only with spears.

The East Indian pig is somewhat smaller and more refined than the European pig. Originating in the East Indies and southeastern Asia, *Sus vittatus* comprises a number of races including the domestic pig of China. This blood is evident in many of our modern domestic breeds.

Because swine generally do not migrate great distances, there are many subclassifications of both *Sus scrofa* and *Sus vittatus*. Authorities generally agree that all of our modern domestic breeds can be traced back to these two species.

Pigs are not native to North America. Columbus introduced them here on his second voyage, which gives some indication of the esteem in which swine were held in Europe at that time. With no natural enemies (except man) the eight original hogs prospered, and records indicate that thirteen years later numbers of them were killing cattle, forcing the settlers to hunt these wild swine with dogs. Hernando De Soto took thirteen hogs on his exploratory journey from the Everglades to the Ozarks. Three years later the herd numbered seven hundred. No doubt those that escaped were the ancestors of the wild pigs found by early settlers in those areas.

By 1790, six million pounds of pork and lard were exported from the colonies. Thus, hogs had an early impact on the agricultural exports of America. However, there was little or no pig farming as such. The animals were virtually wild, roaming the New England countryside at will, foraging what they could. There was no orderly marketing process. The animals were hunted, usually with dogs. In this regard early hog production was more akin to hunting deer and buffalo than to agriculture. But as population increased and agriculture became more established and organized, domestic hog production took hold.

Cincinnati became the first pork-packing center in the United States and by 1850 was known as "Porkopolis." The number of hogs slaughtered in Cincinnati went from 85, 000 in 1833 to 250,000 ten years later, to 360,000 in 1853, and to 606,457 in 1863.

This growth was coupled with the development of what

was then the western frontier. Fifteen bushels of corn could be packed into a pig, the pig packed into a barrel, and the barrel sent over the mountains to feed mankind; and given the transportation of that day, transporting the barrel was more economical than shipping the corn. This system made it feasible to open the agricultural frontier. In a way then, hogs opened the Midwest to agriculture (and ultimately to other manifestations of civilization).

China probably still has the largest pig population in the world, although Chinese agricultural records have not been open. The three other largest pork-producing nations are Brazil with sixty-three million head, the United States with fifty-five million, and West Germany with nineteen million. Of the world's red meat supply, 38 percent is pork—a considerable portion, especially in view of the sanctions against pork by some major religions.

Getting comfortable with the terms

We've been referring to swine, pigs, and hogs interchangeably here, mostly because using but one term gets a little tiresome. Technically, swine is the generic name. In the United States, *pigs* refer to swine weighing under 120 pounds and *hogs* are those weighing more. But in Britain, all are pigs.

The female is a *sow,* the male is a *boar.* A young female that has not yet given birth, however, is known as a *gilt.* A young castrated male is a *barrow.* An old boar that is castrated is a *stag.* (Boar meat has a strong, gamey flavor and aroma, and castrating the sexually active animal several weeks before slaughter improves the taste somewhat.)

Another term you might run into in some areas is *shoat.* This most often refers to a young pig. *Weaners* are young pigs that have recently been weaned, and generally fall in a weight range of twenty-five to forty pounds. *Feeders* can refer to swine from weaning-age on up.

In reading or listening to market reports you'll come

across the term *U.S. No. 1 & 2 butchers,* which refers to grading standards; we'll examine this matter later.

Another fairly common term, and becoming more so, is *SPF.* This stands for specific pathogen-free. An SPF hog is one that has been born by caesarean section in a sterile environment and is therefore free of certain disease-producing organisms, or pathogens.

While we'll be looking at some of these terms in more detail later, and a few others besides, these are the basic ones you're likely to encounter as a new hog farmer. And with this background data behind us, we can now progress to information of a more specific nature.

Chapter 2.
BREEDS

Although only 10 percent of the hogs in the United States are purebred, that 10 percent is important to the entire industry because it provides the seed stock. The average commercial producer (and certainly the average homesteader) will not be raising purebred hogs, but he *will* take an interest in them because his foundation stock, at least ultimately, comes from purebreds.

The more popular breeds

As with other classes of livestock, there is no one best breed, because there can be more variation between families and individuals within a breed than there is between different breeds. However, there are general breed characteristics that make one breed more appealing to certain producers than another. Unlike most other classes of livestock, many of the hog breeds found in the United States today originated here. These breeds enjoy varying popularity in different parts of the country.

The *Duroc* is hardy, prolific, and easily identified by its red color. Oddly enough, the breed is named after a horse. It seems there were strains of red hogs in New Jersey which were known as Jersey Reds. There were also some red swine in New York which were developed by a man who owned a famous stallion named Duroc, so folks in the neighborhood called his swine, Durocs. The Durocs and Jersey Reds were later intermingled, and until quite recently the breed was known as Duroc-Jersey. The breed association is the United Duroc Swine Registry.

The popular *Chester White* originated from a mixture of English swine in Chester County, Pennsylvania, and was

Duroc.

Chester White.

once called the Chester County White. It is known as a medium-type hog, showing up well in most breed averages of desirable characteristics. It is registered by the Chester White Swine Record Association.

There is an offshoot of the Chester White, the *OIC* (formerly Ohio Improved Chester). The OIC is white and has a slightly dished face and slightly drooping ears; most hog men feel this breed needs more development to reach today's meat standard. The registry is the OIC Swine Breeders' Association.

The *Berkshire* is one of the oldest of improved breeds of swine, being mentioned in the late 1700s in England and Scotland. The original stock had some Chinese blood, and was reddish or sandy and sometimes spotted. The first Berkshires arrived in the United States in 1823. The most distinctive feature of Berkshires is their dished face. They have erect ears, are of medium size, and are recorded by

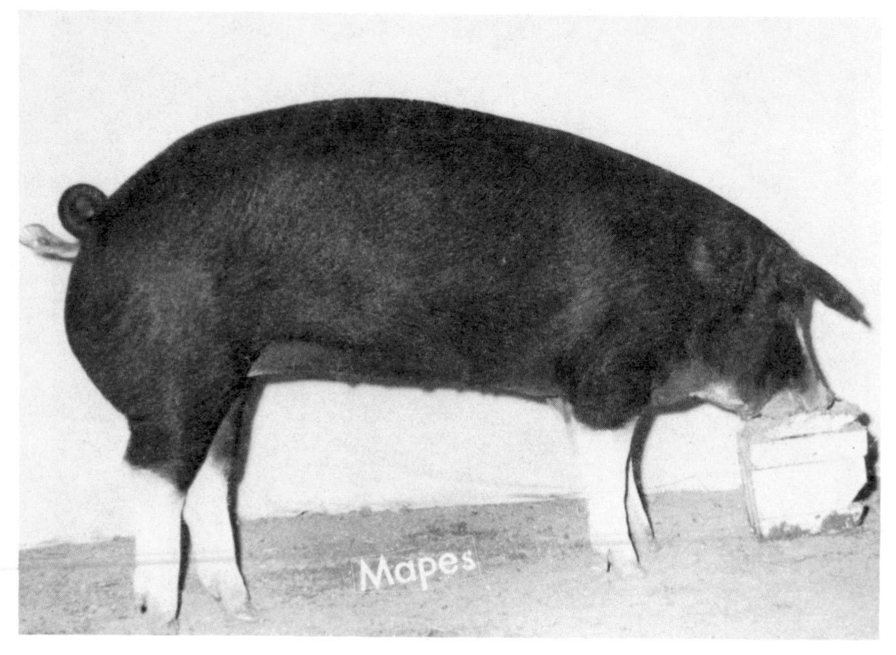

Poland China.

the world's first society for keeping pure a breed of swine —the American Berkshire Association.

The *Poland China* is another American breed (in spite of its name), developed in Ohio in the 1870s. If legend can be believed, the name Poland refers to the fact that the man who sold one of the original animals happened to have been born in Poland. Poland Chinas are generally black and often have white markings, especially on the feet, snout, and tail. Until 1946 there were three registries for Poland Chinas, but at that time they combined into the Poland China Record Association.

Spots, or Spotted Swine, were formerly called Spotted Poland Chinas, and they have a great deal of Poland China blood. One of the breed requirements is that at least 20 percent but not more than 80 percent of the body surface be white. Conformation of the Spotted is very similar to that of the Poland China, though the Spotted is generally a little smaller. This breed was developed chiefly in Indiana, and liberal crossings with Poland Chinas were in evidence until the 1920s.

In contrast with the Spotted, the *Tamworth* is not only one of the oldest breeds of swine, but one of the purest as well, with evidence of pure breeding going back 150 years. They originated in Ireland where they were known as Irish Grazers. Tamworths are known as bacon-type hogs and there is an unusually large proportion of lean meat on them. They are also quite prolific. Due to its extreme bacon-type, the breed has never become very popular in the United States, although it has been used in improving other breeds.

The *Yorkshire* is another English bacon-type breed, and a relatively pure breed with improvement coming from careful selection rather than outcrossing. There are Large and Middle Yorkshires, with the Large predominant in the United States. Breed characteristics include good mother-

Spotted.

Yorkshire.

ing ability (they farrow and raise large litters and are good milkers) and excellent foraging. Their bodies are long and deep with especially long, deep sides, but the hams sometimes lack plumpness and depth. Yorkshires have been gaining in popularity, especially as breeders have been turning away from lard-type hogs since the price of lard fell below the price of live hogs. The registry is the American Yorkshire Club.

Many homesteaders gain familiarity with rabbits before being introduced to hogs, and the *Hampshire* hog reminds many of them of dutch rabbits. The similarity of the belting is striking. Hampshires are an English breed, at least so far as the distinctive black and white color pattern is concerned. However, Hampshires as a breed are considered to have originated in Kentucky in the late 1800s. Characteristics include trim and light jowls, smooth and well-set shoulders, and erect ears. They are trim and free of excessive lardiness, but in the past, Hampshires have tended to lack body length and fullness of hams. They have been noted for their prolificacy, vigor, foraging ability, and meat quality. They are not among the largest breeds at maturity.

The *American Hereford* apparently came about because a group of Iowa and Nebraska hog breeders wanted to develop a pig that looked like a Hereford cow, which once again proves that it is possible to mix business and pleasure. The National Hereford Hog Record Association was formed in 1934, sponsored by the Polled Hereford Cattle Registry Association. The foundation stock included Chester Whites, OIC's, Durocs, and hogs of unknown origin. The chief attraction of Herefords seems to be the distinctive markings, and they aren't very widely distributed. These pigs are smaller than most other breeds and are considered too lardy and heavy shouldered.

In the same year that the Hereford registry was formed, 1934, the first Danish Landrace swine were imported by the

Hampshire.

USDA. The *American Landrace* is white, has good length of body, and is prolific. The American Landrace Association was formed in 1950. While there are more registered American Landrace swine in the United States than Tamworth or Hereford, perhaps their greatest importance lies in their contribution to many of the newer, "inbred" breeds. These include Beltsville No. 1, Lacombe, Maryland No. 1, Minnesota No. 1, and Montana No. 1.

In themselves, breeds are only of minor importance to the commercial hog producer and perhaps of even less importance to the homesteader. Yet, this is not to say that a knowledge of breeds and breed characteristics isn't important. The farmer who farrows sows should be using purebred boars even if the sows are crossed, and the farmer who just buys feeder pigs takes more than a passing interest in the genetic potential of those pigs to do the job he requires of them. That means that even if you buy a single pig and put it in the freezer a mere four months later, somewhere along the line somebody had to pay attention to breeds and breeding.

Choosing a breed

What breed is best? There is no best breed. The homesteader wants good, thrifty, healthy, fast-gaining pigs with desirable meat qualities, and these can be had in nearly any breed (with the reservations mentioned in the breed descriptions above). Anyone farrowing hogs will want the best stock he can afford, and this generally means sticking with the breeds already popular in a particular region. The purebred breeder, involved in showing, registrations, and sales of breeding stock as well as production of commercial pork, will take the most interest in breeds of swine, but even then the choice is highly personal. Good stock must be available, not only for starting out but for infusions of new blood from time to time, and there must be a market for the stock produced. So once again, the breeds the neighbors raise are a prime consideration.

There are certain breed characteristics that have been charted by researchers. While the practical uses of this information are limited, if only because of the great variations among animals of the same breed, the figures do show trends. (The data was collected and compiled by a class in "Breeds of Livestock" at the University of Wisconsin in 1958.)

Average litter size by breed

Berkshire	8.07
Chester White	10.58
Duroc	9.83
Hampshire	9.63
Landrace	8.58
Poland China	8.54
Spotted	10.01
Tamworth	9.00
Yorkshire	11.13

Average loin eye size, carcass length, and backfat thickness by breed

Breed	*Average loin eye size (sq. inches)*	*Average carcass length (inches)*	*Average Backfat (inches)*
Berkshire	3.79	29.5	1.47
Chester White	3.36	29.3	1.65
Crossbred	3.60	29.2	1.78
Duroc	3.31	29.1	1.64
Hampshire	4.13	29.5	1.52
Landrace	3.33	31.1	1.53
Poland China	4.20	28.8	1.54
Spotted	4.06	29.0	1.63
Tamworth	3.35	31.1	1.59
Yorkshire	3.70	30.8	1.60

Dressing percentage, following, is the amount of meat you get from a live animal. A farmer might get 50¢ a pound for live hogs while you pay $2 a pound for bacon, because live pigs have skin, bones, ears, and guts that don't have much market value. We'll be getting into that and other terms in detail later.

Average dressing percentage by breed

Breed	%
Berkshire	71.6%
Chester White	72.9
Duroc	71.0
Hampshire	71.6
Landrace	71.4
Poland China	72.5
Spotted	73.1
Tamworth	70.8
Yorkshire	72.1

These charts show several things. First, no breed excels in every trait—for that matter, there is no breed that is above average in all traits. There isn't even a breed that is in the top three in every trait. There are, however, some breeds that tend to be in the top group more than others. This is still no reason to say these breeds are best, because this denies that there is plenty of room for improvement, and ignores the fact that this type of data isn't available on other important traits, such as feet and legs, underlines, feed conversion, and growth rate.

The individual producer can compare figures from his herd against the averages presented here to see how he stacks up, and to find where there is the most room for improvement. However, these are *averages,* and improvement can come only from using animals that are better than average. As averages are exceeded in a given herd, further improvement becomes more difficult and additional concentration is required just to maintain the same pace.

Chapter 3.
CROSSBREEDING

It can be seen that breeds of swine are really of less importance to pork producers than their type. That is, the color, the shape of the head and face, and erect or drooping ears really have little to do with profit or carcass quality.

Swine type—the ideal having all the characteristics suited to a husbander's purpose—is easily changed by breeders because of the early maturity of swine, their prolificacy, and the resultant short time between generations, as well as genetic factors. A cow reaches sexual maturity at eight to twelve months of age and has a gestation period of nine months, so it takes about a year and one-half for a newborn cow to produce another cow. A hog will reach puberty in about six months and produce not just one but six or seven or eight young, less than four months later. The gilts in that litter likewise could farrow less than a year later. So in theory at least, in the two years or so it takes a cow herd to double in size, a herd of pigs could increase fifty or sixty times! (The original mother would be farrowing again when her daughters are farrowing.)

This is one of the factors that makes it relatively easy for new farmers to get started with hogs. It also makes it easier for knowledgeable breeders to improve hogs, because they can plan matings in terms of months rather than years, and they have far greater numbers of animals to select from.

Type

How does a breeder know what type of hog to breed? Largely through consumer demand. Processors obviously want the kind of meat consumers will pay for, which means the processors are willing to pay more for the kind of hog that will produce that meat. This leaves the farmer with little choice but to breed that type of hog.

Originally hogs were valued for immense size and fattening ability. In the late nineteenth century, early maturity and a thick finish became desirable, resulting in hogs that were smaller and more compact, with very short legs. This was known as a "chuffy" type. As a result of breeding for this type (and neglecting other factors), sows often farrowed twins and triplets, and weight gains of over two hundred pounds were very expensive.

About 1915, big-type strains became the fad—there was a complete turnabout. Breeders went after the exact opposite of the chuffy type. Length of body, plenty of bone, growthiness, and great size became the new ideal. Show ring champions of that day boasted long legs, weak loins, and "cat hams." They were slow in maturing and required great weight in order to attain market finish.

Since the mid-twenties, breeders have sought a balance between the two extremes, leading to a product that makes sense not only for the ultimate consumer of pork products, but for the producer as well. Large litters, fast growth, and quality cuts benefit everyone.

At one time lard was an important pork product, but since the advent of other oils, the demand for and value of lard has decreased. Therefore a lard-type hog fell out of favor. Bacon-type hogs are the opposite of the lard-type. They lay on little fat, a quality enhanced in such countries as Denmark by feeding methods; instead of corn, hogs are fed dairy byproducts, barley, wheat, peas, other grains, and root crops, resulting in a high quality bacon that has earned a reputation in international trade.

American hogs are somewhere between lard and bacon types. In recent years, demand has been affected by our sedentary habits and concern with cholesterol, as many folks now look for leaner cuts. The ideal meaty-type hog of today can reach market finish without excess fat and has good length of body and muscling. The breed has relatively little to do with the attainment of this ideal, as most of the common breeds are similar in these aspects. But *breeding* has much to do with it. One of the most important parts of breeding management is selection.

The ideal hog is a meaty animal with good length of body and heavy bone. It should not be lardy, and therefore should not have excess backfat (see chapter 12). Breeding stock should not be kept unless boars at two hundred pounds have less than 1.25 inches of backfat and two-hundred-pound gilts have less than 1.4 inches. Littermates going to market should have less than 1.6 inches of backfat.

Marketed littermates should have at least four square-inches of loin eye (see chapter 12). Breeding stock should have had a feed conversion ratio of 3.25—that is, it should not have taken more than 325 pounds of feed to produce one hundred pounds of gain from weaning to two hundred pounds.

Keeping the breeding stock from large litters is important, although litter size is not highly heritable. Litters should number at least eight, and preferably ten. Weight at weaning is also important. Pigs from gilts should weigh at least twenty-four pounds at six weeks and thirty-five pounds at eight weeks. Those from sows should weigh twenty-seven pounds or more at six weeks and at least forty pounds at eight weeks.

Gilts should reach two hundred pounds at 165 days of age or sooner, and boars should reach that weight two weeks earlier. Both boars and gilts to be used for breeding should have at least twelve well-developed teat sections. Breeding stock should obviously be free from diseases of any kind and from abnormalities and heritable defects such as cryptorchidism (one or both testicles retained in the body cavity) or hernia.

Some of the more economically important heritable traits of swine include feed conversion, number of pigs raised to market weight per sow per year, growth rate, carcass quality, dressing percentage, and longevity of breeding stock. Furthermore, each of these is determined by many other factors. For example, the number of pigs marketed per year depends on the ovulation rate, fertilization rate, embryonic and fetal survival, milking ability of the sow, motherliness, resistance to disease, number of functioning teats, and so on.

Some economically important characteristics of swine have a low heritability—that is, the traits are not readily passed from parent to offspring. Fortunately, however, some of these traits are enhanced through crossbreeding.

Why crossbreed?

Crossbreeding is basically the mating of two animals of different breeds, such as a Duroc-Chester White cross. In practice, the boar is usually purebred while the sow is a cross. Generally speaking, crossbreeding is frowned upon in livestock circles because once a breed has been "fixed," it has certain predictable characteristics and is said to be homozygous. (There are a few exceptions. For example, Brown Swiss heifers frequently have trouble calving because that breed is known for its large calves. Therefore, first-calf heifers are often bred to Angus bulls, because that breed is known for its small calves. Another example would be the crossing of a Finn sheep with a meat-type sheep. Finns are famous for their multiple births, and just as notorious for their lack of meat. The offspring will not be as meaty as the meaty parent, nor as prolific as the Finn parent, but will be a desirable blend of each.)

Among common farm animals, only sheep and swine are crossbred with any regularity. It has been estimated that 90 percent of the swine in the United States are crossbred.

The case for crossing swine has been well established in several tests, so it is by no means a hit-or-miss breeding method. Meatiness is not greatly improved by crossing, but there are other benefits:

- Crossbred pigs reach market weights an average of seven days sooner than purebreds, on the same amount of feed.
- Crossbred sows average an 11 percent increase in litter size.
- Crossbred pigs suffer a 13 percent lower death rate in the period from birth to weaning.
- Crossbred litters weigh about 6 percent more at weaning, and crossbred sows wean about 15 percent heavier litters than purebred sows.

The above figures were based on 1,700 litters and 8,000 pigs studied at Oklahoma State University, and other tests have shown similar results.

Several interesting facts emerged from this study. For one thing, litter size actually decreased when purebred boars of one breed were mated to purebred sows of another breed. However, when crossbred sows were mated to purebred boars, the litters were consistently 11 percent larger. And only 67 percent of the purebreds survived to weaning, compared with 80 percent of the crossbreds.

It's also interesting that important traits with low heritability are those that respond well to crossbreeding. For example, heritability of carcass length, an important consideration in the analysis of the final product of hog production, is estimated at 60 percent; loin eye area, 48 percent; backfat thickness, 50 percent; and ratio of ham to carcass, 58 percent. Heritability of these traits through crossbreeding is fairly high, and improvement can come about through selection.

However, the number of crossbred pigs farrowed amounts to 15 percent heritability; number weaned, 15 percent; and birth weight only 5 percent. That means that these last items, which are also of obvious economic importance, cannot be insured merely by selection. A sow that farrows and weans a large number of good-sized pigs will not necessarily have daughters that do as well. But as we've just seen, crossbreeding results in an increase of 15 percent in litter size, birth weights are 10 percent heavier, and death losses from birth to weaning are 13 percent lower. This, then, explains the importance of crossbreeding. There is more to be gained in certain areas of economic importance by crossing breeds than there is from selecting parents with the desired traits. An important term here is *heterosis,* or crossbred vigor, which refers to the increased performance of crossbred animals.

It must constantly be stressed that crossbreeding is not haphazard, detracts in no way from the importance of purebred stock, and can in no way be equated with mongrelization.

Systems of crossbreeding

There are various systems of crossbreeding, each of which has its supporters. The simplest involves crossing two different breeds, or a purebred male of one breed with high-grade females of another. ("Grade" here refers to animals that display many characteristics of a recognized breed but are not purebred.) Sow replacement can be a problem with this system because the sows will have to be bred to a boar of their own breed for replacement gilts.

A two-breed rotation uses boars of two different breeds

in alternate generations. Crossbred sows are mated to boars of the same breed as the grandsire (grandfather) on the dam's (female parent's) side.

In a three-breed rotation, crossbred sows are bred to a purebred boar of a third breed, whose gilt offspring are then bred to a boar of the first breed.

Other systems can use more breeds, or in certain sequences, and these become similar to type-crossing, which involves crossbred sows where the choice of a boar depends more on type than on breed. There is no set rotation. In practice, there appears to be little advantage in using more than three breeds.

It's important to emphasize that crossbreeding does not replace selection, if only because the increased performance due to heterosis is a one-shot affair. The effects of hybrid vigor do not accumulate over time as do the factors which respond to selection. Only genetically superior stock, of both the pure and crossed lines, should be used for breeding. This again demonstrates that when we speak of crossbreeding the way we do here, we're not talking about the haphazard creation of mongrels.

In modern commercial hog crossbreeding, only meat-type animals are used, rather than contrasting types. This varies somewhat from usual crossbreeding in other classes of stock, for which the goal is to use sharply contrasting types in order to get a wide variability which will provide the greatest possible selection. In that case, animals with the desired combination of two types are selected, and the traits are fixed by inbreeding. In the hands of skilled geneticists, this is how new breeds are developed.

But the commercial hog man isn't interested in coming up with something new or different. He wants large litters of fast-growing pigs that develop into top market hogs on a minimum amount of feed, and he wants them to be uniform. Therefore a knowledge of type is important, as is a knowledge of the traits of major breeds and careful attention to selection.

Chapter 4.

MANAGEMENT METHODS

The key factor in hog production is management. A good manager can do better with poor hogs than a poor manager can do with good hogs, and in this regard, management is even more important than the animals themselves. Moreover, management to a large degree *determines* the quality of the animals, because it also includes selection and similar factors.

Management is the art of handling or directing something with skill—the judicious use of means to achieve a desired goal. Hog management is the art of skillfully raising swine, and includes procuring stock, housing, feeding, controlling disease—in brief, everything that leads to the desired goal, which is a market-weight hog.

Because management is an art, there can be no set rules and methods. Different management methods work with varying degrees of success for different people. This means there is room for a certain amount of individuality and creativity. Even aside from that consideration, hog raising is chock-full of options.

The matter of scale

First, of course, you must decide if you are going to raise one or two pigs strictly for home use, or go commercial. Going commercial can mean turning out thousands of hogs a year, or it might mean just raising a couple for some friends or relatives. On the smaller side, hog raising can utilize homestead management methods and still be commercial. It's easy to tell when a homestead operation becomes a homestead-commercial operation, but it isn't so easy to say when a homestead-commercial hog farm becomes strictly a commercial venture. The two goals—homestead meat and cash—melt together into an indistinct blur, and there is sometimes even less distinction in management methods.

The goals of the homesteader are first, to provide meat for his own table, and in conjunction with that, to produce it as efficiently as possible from the standpoint of both labor and capital. In addition, it must be a quality product, which is generally taken to mean that it should not include antibiotics or other chemicals as well as meeting more common commercial quality requirements. Lesser homestead goals might include the utilization of surplus and waste, and the production of manure for fertilizer.

Some of these goals spill over into commercial hog farming of course, but the main reason for raising hogs commercially is to earn money. While some management methods would fit either homesteads or commercial farms, the basic difference between goals is more likely to dictate differences in management.

The homesteader can expect to harvest 135 pounds of retail cuts from the average market hog. Since average per capita consumption of pork in the United States has been 72 pounds, one hog will supply two people for a year, and the average family would only need two or three per year. (It would be hard to prove, but there's a good chance that people who raise their own pork consume more than the national average—if only because it's such doggone good eating!)

The homestead that produces pork strictly for its own use will not be involved in farrowing. A sow should produce six to eight pigs a year at least, which would keep twelve to sixteen people in pork. Other management factors that enter into this include the cost of a boar and his maintenance, which would be prohibitive for one sow because the cost of the boar's upkeep must be charged against the pork produced. (Boars can sometimes be borrowed from neighbors for just the cost of the feed.) An even more serious consideration is the fact that sows often do *not* produce six to eight offspring: many come up with two or three, and some experts have estimated that 10 to 15 percent of the female swine population is sterile, which of course means zero piglets. The family that depends on its own hog pen for its BLT sandwiches will be much more assured of a supply of bacon by purchasing weaned feeder pigs.

Most farrowing operations do not sell pigs one at a time, but by the truckload. However, just because a farmer looks at you a little strangely when you drive up and ask to buy a single pig, doesn't mean he'll refuse to sell you one. You can locate farmers who raise pigs by watching ads in local newspapers and farm publications, by asking vets and feed dealers, or just by driving around the countryside.

The cost

Prices for weaned pigs vary as drastically as prices for market hogs, and the two are usually connected. One common rule of thumb for setting feeder pig prices is to multiply the weight of the pig up to 40 pounds by 1.6, then multiply that by the current market price of butcher hogs. You can get the going price from radio farm market reports, from newspapers that print livestock market news, or by phoning a buyer or packer. If the pig weighs from 40 to 60 pounds, use 1.5 for the factor instead of 1.6.

To illustrate, the market today was 50¢. Multiply the weight of the pig (let's say 40 pounds) times 1.6, which is 64. That, multiplied by the market price (50¢ a pound) is $32. As with all rules of thumb, this is only a guide, but at least it offers a starting point. Many other factors might be involved: the outlook for grain prices can affect the prospects for feeding hogs; barrows might be worth slightly more than gilts because they have a slightly better rate of gain; quality is certainly a factor; or the farmer might consider that he's spending more time selling you one pig (and answering your questions) than he'd spend selling a truckful to another buyer. On the other hand, maybe the answers you get to those questions will be worth more than the extra cost of the pig!

In this kind of trading the price can also depend on how badly the seller wants to sell and the buyer wants to buy, as well as the knowledge and integrity of the two.

When you buy a feeder pig you are paying for the cost and the upkeep of the sow and the boar plus the feed that the young pig has consumed. This includes depreciation on housing and equipment, as well as a return on labor, management, and investment for the farmer. While none of these might be of paramount importance to the homesteader who just wants "a pig," the person who wants to be a good manager will want to take advantage of the helpful information of anyone who furnishes the foundation for his homestead hog enterprise.

You can't make a silk purse out of a sow's ear, and you can't expect to raise a pig successfully if the animal just doesn't have what it takes. The prime factor here is genetic potential. Many homesteaders get stuck with, or even choose to start with, a runt. If the price of pigs is high and a runt can be obtained very cheaply, it might work. But more often, a runt will remain a runt and will require much more feed than a good sound pig would. That makes the pork very expensive, even if the farmer were willing to *give* you the runt. What's more, the quality of the meat might not be up to par, making the runt a decided disadvantage. Also, it is prone to other problems.

Picking out a pig

The homesteader wants a good pig, from a good farmer. A feeder pig should be weaned, of course, and if a male, castrated. If your pig comes from an operation of any size, the manager will also have performed other tasks. The pig will probably have had iron shots and possibly others. It might have been wormed. The needle teeth will probably have been clipped, mainly for the comfort and protection of the sow. The ears might have been notched as an identification measure. The tail might have been docked to prevent tail biting, which of course is no problem if you only have one pig and is of little concern even in larger operations if the pigs have sufficient room. (It should also be noted that farmers who dock tails are beginning to see cases of ear biting, which is a good case for eliminating the cause of a problem, not just the symptoms.)

Moving any animal creates a stress situation which demands good management. The pig is subject to unaccustomed handling, transportation, different surroundings, different feed and water—and for the single pig on the homestead, the strange condition of loneliness. The first order of business when dealing with a new animal is to make it as comfortable as possible and allow it to become adjusted.

Unlike a commercial farmer, you might have picked out your first pig the way some people pick out a puppy: he was cute, or friendly, or had a white spot on his forehead that looked like a star. But what the creature needs at this point is not to be petted by the kids or hassled by the dog or hovered over by the homesteader, but to be left alone. Provide clean bedding; clean water and food; protection from the sun, wind, and rain; and peace and quiet.

Even if you don't have any experience with pigs, you should have been able to tell by looking at a pen of several that the one you picked showed good health. A healthy animal is brisk and alert, with a bright eye and good skin and coat. If you made a good choice, clean water, good nutrition, and a clean place to sleep will help the little pig maintain that health.

While most commercial hog men follow a rigid regimen for starting feeder pigs including medications in water and feed and drugs for both internal and external parasites, the homesteader who truly wants to eat food that has not been contaminated by antibiotics and other chemicals is by no means courting disaster if these products are not used. After all, these drugs are a relatively recent innovation, and pigs were raised for thousands of years without them.

Medication and synthetic feeds

Swine medication—both as a preventative and as a growth stimulant—has become so commonplace today that this aspect deserves further attention. It touches on many of the differences between homestead and commercial management.

The homesteader wants good food. The farmer wants profits. In most cases today, this means the homesteader will want to raise pigs organically and the farmer will want to use toxic technology; because most of us tend to think in terms of black and white, organically grown means good food and toxic technology means profits. (An interesting corollary would be that farmers cannot produce good food.)

But look at the vicious circle this line of reasoning results in. We might say that the organic homesteader can raise pigs organically because he raises them organically, and the farmer who uses technology needs an ever-increasing amount of technology. Let me explain.

The homesteader raises one or a few pigs at a time. The commercial producer raises many, and very often under crowded conditions, because building and equipment expense is reflected in profits or lack of profits. So at the very beginning homestead pigs are less crowded, less stressed, less likely to encounter disease or injury or parasites, and more likely to get individual attention from the caretaker. Under these conditions hogs have less *need* for the generally accepted arsenal of medications. Thus we might say that organic homesteaders can raise hogs naturally because they raise them naturally!

In addition to housing differences, on the well-run homestead sanitation is extremely important, not only because sickly and filthy animals do not contribute to the goal of "good food," but because the bedding and manure are extremely valuable byproducts for other facets of the organic homestead. Good sanitation shows up in animal health and well-being, and again lessens the need for medications.

The homestead hog is much more likely than the commercial one to enjoy a varied diet which, while not necessarily a balanced diet, would seem to be a more natural diet. While it might be true that most animals eat better than some humans, there is still something disturbing about pigs that are fed nothing but manufactured supplements and corn ("synthetic" corn at that, if it's grown on synthetically fertilized soil). The experts with the test tubes (and the grants from the chemical companies) disagree of course, but not only have they not convinced many dirt farmers, the organic or natural viewpoint is gaining new converts daily. We really know little about nutrition, and some of the best examples of this are found in hog husbandry. Even scientists admit that feeding milk to hogs lessens problems with worms, but no one knows why. They also agree that giving swine a clump of sod or a bucketful of uncontaminated soil or letting them root in a clean pasture provides not only the minerals and trace elements which are incorporated into commercially mixed feeds and premixes, but other, unidentified elements as well. Under these circumstances it's hard to be convinced that carefully formulated commercial rations are as "complete" as their vendors like to claim. In the homestead situation there is bound to be surplus milk during the spring and summer, or whey or skim. There will be comfrey and garden produce and fruit. It's common knowledge that pigs are used in research because their digestive systems are so similar to man's, but do you think you'd be healthy, or happy, eating nothing but corn and chemicals?

So, the more you raise pigs under "natural" conditions, the more "natural" you can afford to become. And conversely, the more you rely on technology, the more dependence you have to place on technology in the future.

Now, we must back off a bit and look at this a little more realistically. Good hog farmers will be quick to point out that all of this is overdrawn and theoretical. Modern commercial hog facilities *are* sanitary, and perhaps more so than even the best homestead setups in some cases. Commercial producers will point out that their feeds, tested for protein content and formulated by trained nutritionists, are far preferable to the tidbits fed by homesteaders with no knowledge of nutrition either from the standpoint of what they're feeding or the pig's needs. And of course they will mention that routine medication results in fewer losses and faster and more efficient gains, which means more and cheaper food for the consumer.

This is true insofar as there are some very good commercial hog producers and some very poor homestead producers. Conceding that, the fact remains that in pursuing ideals, the homesteader who is truly serious about living organically need have no qualms about raising hogs organically.

Whether or not you decide to medicate your healthy hog, the number-one job from here on in is providing feed, clean water and shelter, and maintaining sanitation. Of course, even these simple matters require knowledge and involve options, most of which will be covered in some detail when we talk about housing and equipment. However, there are a few management methods uniquely suited to the homestead.

Homestead methods

One rather unusual practice favored by some people is tethering the pig. A special harness is required, but most hogs readily adapt to being tethered, and it makes a great conversation piece. Most visitors will be extremely wary of the watch-pig staked out in your yard! This system involves the labor of moving the pig from the pen in the morning and back at night, and the animal should always have shade available as well as water. But there are several advantages. By moving the stake regularly, the pig will always be on clean ground. He will gain a certain amount of nourishment (depending on the location and the time of the year) and will fertilize the area at the same time, with no effort on your part. He will also "plow" the area, so make sure it's a garden or future garden, not the lawn.

This plowing habit can be used to good advantage without the bother of the tether and in a more methodical fashion by using a small moveable pen. Feeder, waterer, and shade can be incorporated right into the pen. The pen should be large and heavy enough to contain the pig, but small and light enough to be moved. Your imagination and the resources at hand are the deciding factors in planning the construction of such a pen.

The idea is to start at one corner of the area to be worked up and let the pig root to its heart's content within the confines of the pen. He will eat the vegetation, turn up and devour roots and grubs, and do just a dandy job of turning that piece of ground into a potential garden, including fertilizing it. The pen is then moved down the line a notch.

Pigs are also great gleaners and after harvest can be turned into gardens and fields to scrounge whatever's left. In larger areas the plowing and fertilizing effect will be less, but the pig will derive more nourishment from the foraging.

If fencing is required to keep hogs in the garden or field, one of the best temporary answers is an electric fence about nine inches from the ground. The pig will have to be trained to respect it by your first running an electric fence around a small area with a board fence or stock panels outside it, but this can be worth the effort. The wire and posts for such an electric fence are inexpensive and easily moved.

Pigs can also be run with cattle, and will pick up nourishment from the feed that passes through the cattle's digestive systems. If whole shelled corn is fed to calves, enough will pass through unscathed to provide one-third of the feed requirements for a hundred-pound pig, so with three calves you could keep one pig at no cost except for protein supplement.

Because yearling cattle eat more and waste more, only two steers are needed for one pig. But cattle make better use of ground or rolled grain, and the number of pigs that can follow cattle on a ground ration is reduced by half.

Pigs can injure heifers by rooting at the vulva when the cattle are lying down, and we've had young pigs stand on their hind legs to nurse our Jersey cow!

Some of these systems would meet the requirements of commercial producers; others would not. But the point is that the person with one pig has a great many options in the way that pig is fed and handled. This is not to say that the commercial producer doesn't have options! There are probably more ways to be a hog farmer than anything else in agriculture.

We've been dealing with the strict homesteader who produces pork only for home use. And we mentioned that the homesteader can raise a few extras for sale without becoming a real farmer, or he can raise lots of pigs and become seriously commercial. But even this is only the beginning.

Raising hogs for market

The commercial producer can buy feeder pigs and raise them to butcher weight. He can maintain his own sow herd, and raise the feeder pigs to butcher weight. Or he can keep sows and sell feeder pigs to other farmers and homesteaders. There are also those who specialize in purebred stock. These people pay a great deal of attention to breeding, and they participate in shows and registry associations and in time build up a reputation for fine hogs; as a result they command premium prices for breeding stock as well as selling culls and excess animals on the market.

There are still more options within each of these categories. Hogs can be fed grain grown on the farm, or the grain can be purchased. If sixty sows constitute a full-time operation, the farmer can farrow anywhere from one to sixty and have a full-time job or a one-sixtieth full-time job or anywhere between.

Your choice depends primarily on what scale of operation you want. That might be just a small hog enterprise that can be managed before and after a regular job in town, or a small operation conducted in conjunction with some other type of farming. Someone else might be interested in farming full time but might lack the experience and capital required to break into large-scale agriculture. Such a person might start out with a few sows, gain experience, build up the herd, and work his way into a full-time position. Still another might want hogs to convert a temporary glut of grain or other products into saleable pork.

This is all part of management. It all involves making decisions on what the individual wants to do, and how, and why. Since hog raising is among the most flexible of farm enterprises, it can be molded to fit into a wide variety of situations. But this means that management—the art of directing something with skill—is of prime importance. There are basic rules, to be sure, but there is so much room for individual initiative, for personal decisions, and for innovation, that only the knowledgeable, above-average manager can reasonably expect to succeed.

Chapter 5.
HOUSING

Housing represents anywhere from 8 to 20 percent of the cost of producing pork. Most of the rest goes to feed. But since housing not only affects the comfort and well-being of the pigs—to some degree it determines the amount of feed consumed and the amount of labor involved—its importance is probably greater than the cost factor alone would indicate.

The homesteader might not give a great deal of thought or planning to housing. If only one or two animals are involved, the question is not so likely to be, "What type of housing is desirable?" as "Where can we keep the pigs?"

Many a homesteader and small farmer keeps the pigs in the hog house, which was probably built a generation or so ago and which more likely than not has been vacant for years. Hogs are also kept in old chicken houses, dairy barns, other assorted buildings, and even garages. Not many people who plan to raise a couple of hogs are going to spend hundreds of dollars on facilities, and in most cases it isn't necessary.

Because of the wide variety of situations, of buildings available and their conditions, of climates, and even personal factors, a comprehensive discussion of homestead hog housing is obviously impossible. If you live in the South you won't have to worry about blizzards and freezing temperatures, but if you raise pigs just in the summer you won't have to worry about that in the North either. If you have a serviceable building, your situation is different from someone who has to rig something up. Of course, there are general specifications that will help you do a good job of providing shelter for your pigs no matter where you start.

Portable housing

There are two basic types of hog shelters, portable and permanent. Portable houses are used with pasture systems and may be for gestating swine, farrowing, or feeding. As the name implies, they can be moved where they're needed. They vary in size and style with the use intended, materials available, and the carpentry capabilities of the farmer. The A-frame is popular because of its ease of construction and efficient use of materials, but these houses may also be the shed type, gambrel, mansard, ridge, or a combination. They can have hinged sides that can be opened in fine weather, or even be three-sided in mild climates.

Portable houses have several advantages. They are relatively simple and inexpensive to build. They are flexible. They provide the hogs with more isolation, segregation, and exercise, and they make it easy to rotate pastures.

On the other hand they require more labor, they're harder to keep warm in northern climates, and the hogs

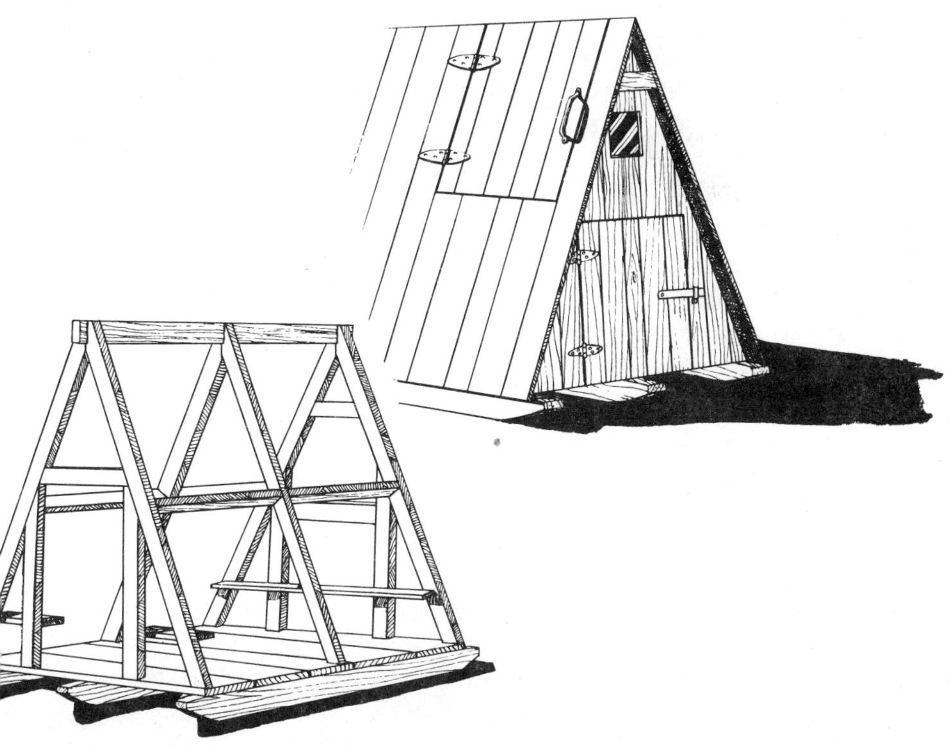

A-frame portable hog house.

usually get less supervision. The latter point is particularly important if sows are farrowed in these houses.

Permanent housing

A so-called central house is a permanent multipurpose house, and as such is something of a stepping stone between pasture methods of management and the more popular confinement systems. Central houses can serve any housing function, from boars and gestating sows through to market hogs. They are often used in conjunction with portable housing. Central houses consist of a series of pens in which different classes of hogs may be kept. There are often sun porches or lots on the south side of these buildings and an aisle inside the north side so the farmer can care for the animals without entering each pen. However, in units of ten or more, there are often pens in two rows, with an aisle between.

Confinement housing

Commercial hog producers are increasingly turning to confinement housing, involving permanent houses designed for specific needs. These are usually more expensive than any of the multipurpose shelters, but the expense is justified by greater efficiency. Labor is reduced; more economic returns are possible from the land available; more pounds of pork per square foot can be crammed into these facilities; and heating, cooling, and sanitation are often improved through design. While there are as many types of confinement systems as there are farmers, these systems are the least applicable to the average homestead. They do have definite possibilities for the serious hog raiser, however, even on a small scale. What's more, certain details are sometimes adopted by even small homesteaders.

Confinement systems are designed for gestating sows, and boars; as farrowing houses; nurseries; growing houses; and finishing houses. Large, integrated operations move the sows into the farrowing house when their litters are due. The special facilities make it easy to care for the sows during parturition and to attend to the newborn pigs. Even more importantly, such facilities offer special protection to the baby pigs. The sows may be in farrowing crates to lessen the possibility of babies being crushed by a mother lying on them; and the temperature can be maintained between 80° and 90° F. in the brooder area. If crates are not used, there will be special guardrails inside the pen to protect the babies from being crushed. The pens may be equipped with creep areas where the babies can keep warm and begin eating specially mixed feeds.

After weaning (which can be a matter of days or up to two months, depending upon the experience and the ideas of the farmer), the pigs can be moved to the nursery, and from there to the growing house, and finally to the finishing house. The economies in such a system result from better use of floor space, since small pigs naturally take up less room than those approaching market weight. From weaning to 75 pounds, a pig in confinement housing requires four square feet of space. From 75 to 125 pounds, from six to seven square feet is recommended, and for pigs over 125 pounds, nine square feet of pen space should be allowed.

For the smaller pigs, six per linear foot of feeder space and twenty to twenty-five per watering cup is the recommendation; for those weighing 75 to 125 pounds, four per linear foot of feeder space and twenty to twenty-five per watering cup; and for heavier hogs, three to four per foot of feeder and ten to fifteen per waterer.

In addition to a more efficient use of floor space and more places at feeders and waterers, this system also makes possible a gradual reduction in temperature and makes it easier to provide each class of hog with the best and least expensive nutrition.

Some confinement systems, especially on smaller farms, combine two or even all three of these functions. There may be a nursery and a growing-finishing house, or the weaned pigs may stay in the same pen until marketing.

The goal of confinement housing is efficiency. It can be seen that many labor-saving and pig-saving aids can be incorporated into these structures, which means they are generally expensive. When you have automatic watering, automatic feeding, and automatic manure removal through slatted floors, you are obviously substituting capital for labor. For this to be efficient, the pigs must be kept growing and moving through the pipeline, and the facilities must be kept full. Perhaps this distinguishes modern commercial hog raising from homestead hog raising more than anything else. Like the farmer of yore, the average homesteader can still afford to say "What's time to a hog?" to which the modern farmer replies, "Time is money."

Such systems, especially fully integrated ones, require a great deal of planning and coordination to keep all sections going full tilt. Breedings are planned well in advance to insure a smooth and steady flow of pigs through the various pens. Sows that have weaned their pigs go back to the gestating house.

It was long felt that even if other pigs could thrive in confinement, pregnant sows needed fresh air and exercise. Today, even gestating sows are kept in confinement, in loose stalls, individual pens, or tie stalls. (It's been said that sows have been tethered in tie stalls in Europe for years, although the practice is relatively new in this country.) When farrowing time comes, the cycle starts all over again.

Which of the above housing options you choose depends on many factors, including how many hogs you intend to raise; how committed you are to raising hogs; your short-range and long-range goals and management plans; whether you prefer to be labor intensive or capital intensive; and so on. But no matter what system you pick, there are several general ideas to bear in mind.

Location

Location is important, but again it depends on the type of operation you envision. The one-hog homestead without power equipment would do well to locate the hog pen somewhere near the garden and the compost bins to save strain and steps. It's seldom a good idea to locate swine facilities any closer to the house than necessary, especially upwind, but it's also important to consider the great quantities of water that will have to be hauled where automatic watering isn't available. Likewise, feed storage might be a consideration.

Drainage is important in locating any building, and especially a hog building. Although a hog will enjoy wallowing in the mire on occasion, he won't enjoy living in it constantly during rainy spells any more than you'll enjoy working in it, and of course sanitation and fertilizer recovery are severely hampered by such conditions.

Consider the possibility of future expansion, and leave room. This is probably of more importance with the larger installations, but it's worth remembering in smaller homestead situations. It's much easier to plan ahead. With or without expansion plans, leave room to maneuver trucks and machinery around the pens. This can apply to one-hog operations as well as to very large ones. I raised my first hogs in a small pen in the back yard. I hauled feed back there in hundred-pound bags and did the butchering myself right on the spot. But one spring I became involved in a number of other projects and decided to take a pig to a slaughterhouse. The only trouble was, I couldn't get my truck through the orchard, shade trees, flower beds, and berry bushes that so effectively hid the hog pen from the road. I felt very much like the fellow who built a boat in the basement and couldn't get it out.

Any livestock shelter is best located out of prevailing winds and with a southern exposure if possible, but if you live in a warm climate or raise a pig just during the summer, shade might be of considerable importance.

Construction

After location, the next most important consideration is construction. Little pigs can squeak through very small openings; large pigs can tear down even substantial walls and fences; and pigs of any size can root and dig their way under gates and fences. The nursery-growing-finishing house, especially, must be all things to all pigs.

Most of the construction details are unimportant. The shelter itself can be constructed of wood, stone, brick, cement block, concrete, metal—in brief, anything that the homesteader has available or can afford to buy, and feels comfortable working with. I once built a very workable small pen out of logs, and another person built a pen from heavy shipping pallets obtained from a trucking company.

Whatever material is used and however it's put together, it must be sturdy. This brings up hog psychology, an under-

A simple alternative to a driving hurdle is to put a basket or bucket over the pig's head and back it to where you want it to go.

standing of which will make life on the pig farm much happier. You can't fight a hog, or a hog's nature. Rather, the smart farmer will strive to understand his animal's behavior and work around that, or even use it to his own advantage. Here's an example of the latter. Many people have heard that pigs are very clean animals and always leave their manure in the same spot. But when they get a pig, they complain that that's an old wives' tale: their pig messes up the whole pen, including the feed trough.

The answer to this is simply that their pen is the wrong size or shape. The pig isn't thinking in terms of a bathroom. It chooses a sleeping place that is out of drafts and rain. Then when it gets up, it invariably moves about ten to twelve feet before defecating. If the pen is twelve by twelve feet, the hog can move in any direction. But if the farmer has studied his pig psychology, the pen won't be square, but rather long and narrow, say six or eight feet by fifteen. Then the pig can move in only one direction and house-cleaning is much simpler.

A hog loves to have its back rubbed and if you aren't around to do it, it'll use any obstruction available to rub itself: a fence post, gate, feeder, doorway, or building corner. When a hog gets to weighing two hundred pounds, that puts a lot of pressure on your building and equipment. They must be built to withstand it, and if such rubbing places can be eliminated altogether, so much the better. This can be further resolved by providing a special post for back rubs to keep the pig happy and occupied.

Chewing and rooting are other natural pastimes of a hog. A good-sized one can chew right through board fences and wooden walls in a very short time. But it has to be able to

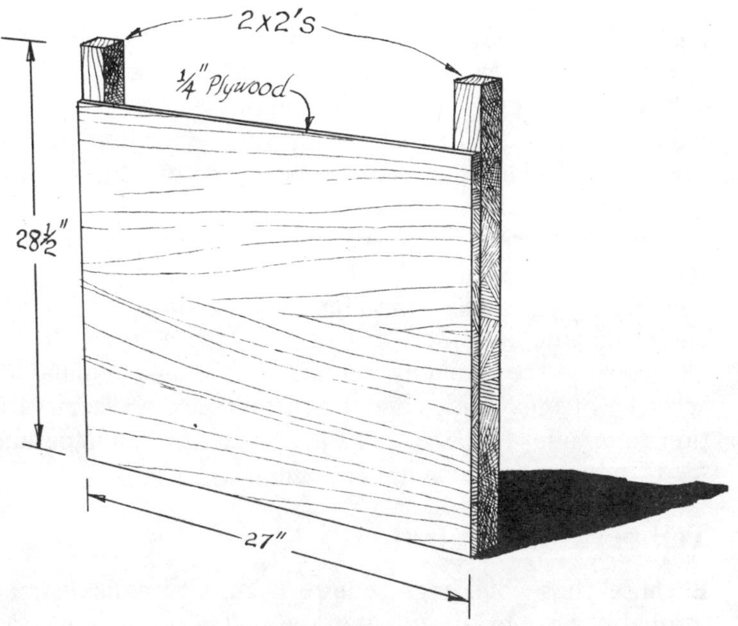

A driving hurdle is very useful for moving hogs. Use either boards or plywood.

get a start, and there are several ways to foil it. Boards can be fit tightly together. And vertical boards are much harder to chew on than horizontal boards. Rooting under fences can be prevented by burying the bottom of the fence; by weighting it down with heavy logs; running an electric fence just inside; or, in the case of a sow or boar at least, ringing the animal's nose.

A pig may like to stand with its front feet on fences and gates, which can get to be a strain on the structure when the animal weighs a couple hundred pounds. It should be built sturdily enough to withstand the pressure. In the case of farrowing crates where a sow with this habit can be a real problem, tethering is a good preventative. A chain anchored about eight inches in front of the feeder is attached to the sow's collar and prevents standing with the front feet on the crate's sides.

If other specific problems arise, either because of your particular conditions or because you have a weird pig, the best way to cope is to watch the animal and learn its habits, and then show you're smarter than the pig by channeling that habit into some less destructive, or even helpful, area.

While much of the information in this chapter might be of interest regardless of the size of your operation, it's obvious that the larger and more specialized buildings will require a great deal more detail than can be provided here.

I can't imagine anyone seriously contemplating big-time hog raising without at least some experience, and most bankers who are called on to finance such projects probably can't imagine it either. But even with a great deal of experience, these highly technical systems demand the services of universities and extension agents, of construction companies that specialize in hog facilities, and perhaps feed companies and equipment manufacturers.

Temperature control

Because these buildings require such a great investment (probably close to 20 percent of the cost of producing pork), they must be capable of being used year-round. That implies heat in the winter and almost year-round in farrowing pens in most locations; cooling in hot periods; and most importantly, ventilation.

Hogs can stand cold better than heat but they're not really built for either, and performance will suffer under extremes of temperature and humidity. Hogs have few sweat glands and not much hair. A hundred-pound pig should require about 2.5 pounds of feed to gain an additional pound of body weight when the temperature is 70° F. But if the temperature is 90°, that same pound of gain will require 4.7 pounds of feed, and at 100°, 7.5 pounds of feed. Similarly, if the temperature drops to 60°, it will take 3.2 pounds of feed; at 50°, 4.1 pounds; and at 40°, 5.3 pounds. The effects of temperature on costs are obvious, especially if the amount of feed required to produce a market hog doubles, and more especially if that doubling is applied to hundreds or even thousands of animals.

The "comfort range" for various classes of pigs is as follows:

Farrowing house (sows)	60°-70° F.
Brooder area (baby pigs)	80°-90°
Nursery (sows)	60°-65°
Baby pigs, 10-50 pounds	70°-75°
Growing house, 50-125 pounds	65°-70°
Finishing house, 125-220 pounds	60°-65°
Gestation house	50°-60°

Maintaining these temperatures (or anything even close to them) will require heat in the winter and, for some classes, cooling in the summer. In the northern United States a heater with an output of 2,500 BTU per hour is recommended for a sow and litter. But since location, type of building, insulation, ventilation, and other factors are involved, each unit is individually designed by heating and cooling experts for maximum efficiency.

There are several common types of cooling. One system

used is regular air conditioning, but ducted to be directed at the sow's head in the farrowing crate. For market hogs or other swine in pens, foggers which spray a fine mist on the animals are common. But again, these are not homestead setups because of the cost in relation to the value returned, and the commercial producer who is making a sizeable investment will want to seek expert help and advice in detail.

The same is true of ventilation, or the changing of air, especially in closed buildings. Ventilation in both winter and summer is important in these structures, and it's a complex and variable process.

The importance of insulation in a building that is cooled or heated can readily be understood in these days of energy consciousness. Even without mechanical heating or cooling, insulation can keep body heat in during the winter and solar heat out during the summer. A one-day-old pig produces 33 BTU/hour; a ten-pound pig, 200 BTU/hour; one hundred pounds, 520; two hundred pounds, 800; and a three-hundred-pound hog will produce 1,025 BTU/hour. Most of this body heat is lost, especially without insulation. And according to agricultural engineers, the heat loss by ventilation sufficient to remove moisture just from the hogs' breathing is assumed to be 60 percent.

Because humidity is so closely related to temperature in its effects on livestock, relative humidity should range between 70 and 80 percent in the hog house. This is not only related to ventilation, but it also points up the need for vapor barriers. Hog houses are damp—from respiration, from urine and feces, from waterers and wet bedding. A vapor barrier of four-mil plastic film or other suitable material placed on the warm side of the insulation (i.e., facing the inside of the house) will help eliminate that frost-on-the-wall problem.

While the profit-oriented hog producer will need to or at least want to make use of as much technology as possible (through trained technicians making specific recommendations for each specific installation), the one-hog homesteader can only concentrate on keeping the pig as warm as possible during the winter and as cool as possible during the summer. It just isn't economically feasible to go to a lot of trouble and expense to save a few dollars worth of feed.

In the event of high summer heat, especially when coupled with high humidity, certain precautions should be taken. The first rule is to not move or otherwise stress livestock of any kind under those conditions—and especially pigs. Provide shade and fresh water. If your waterer is such that the hogs can tip it over and make a wallow of sorts, they'll do it. But the wallow won't be very large or wet, and then they won't have any water to drink, so it's a much better idea for you to make a wallow for them. Or, if you have a hose with a nozzle that adjusts to a fine spray and a faucet close enough to the hog house, you can not only rig up a fogger but create a wallow at the same time. Note that wallows should not be in the shade.

In even moderately larger operations, a ventilation system should be considered even if it's not designed by an engineer. It would be helpful to know that when the temperature is above 70° F., a 125-pound hog needs an air exchange of seventy-five cubic feet per minute (as well as to learn the needs of other size pigs and at other temperatures). An engineer would also work with the size and location of the building and other factors to determine the size, location, and operating speed of the fan. Inlets must be carefully designed, and there are many other considerations. But in the context we're discussing here, almost any kind of ventilation would be better than none at all. A homesteader doesn't consult a refrigeration engineer when he sets a fan in a window on a sultry August day, and that's just about the situation we have here. It isn't efficient but it can make an animal more comfortable, and in the case of a pig in extremely hot and humid weather, it could save the animal's life.

The average homesteader does not maintain hogs in total confinement, and if the buildings can be opened, and foggers or a wallow provided, summer heat is no insurmountable obstacle to raising pork.

Winter may be a more difficult time. Certainly there is little point in trying to farrow hogs in cool or cold weather without supplemental heat. And while market hogs will gain more slowly and expensively, at least they can be maintained with some degree of comfort with little difficulty. Buildings should be insulated and draft-free. But if they're not mechanically ventilated, they should not be closed tight because of the problems moisture will cause. Animals of all kinds face more problems when their owners try to keep them warm by closing them in tight buildings without ventilation; they'll be more comfortable, and healthier, in a colder environment with adequate ventilation.

Wind protection is highly important in winter. Insulation is helpful. Some farmers provide lower ceilings in winter to help hold in heat, and these low roofs can be stacked high with straw for added protection. Deep bedding is good too, as pigs will burrow into large stacks of straw in cold weather to keep warm.

Another possibility that will interest many homesteaders who are constructing a new hog building is to partially bury the structure, preferably in a hillside. You know how cool a cellar is in summer! Some of that effect can be captured in such a building. And of course shelter from winds, and heat from within the earth, would serve to keep it warmer in winter. Experimentation with this type of construction seems to be confined to individuals, and to individuals with imagination and a concern for energy conservation and other matters of ecology. Probably those are the kinds of people who are reading this book.

Along the same line, several people have been working with solar heating of barns in general and hog pens in particular. One system we heard about consisted of a window, an air space, and a cement block wall covered with black plastic. According to the person who built it, the cement block wall stored up heat during the day and appreciably raised the temperature of the shelter.

None of these innovations has been proven, and there are no plans or design details available. Nevertheless, with the number of people becoming interested in this facet of ecology and with the new ideas and products being generated since the energy crisis, it would seem that these uncharted areas offer great potential for the backyard pig farmer—and with some experience and refinement, for larger operators as well.

Inexpensive housing ideas for one to three pigs

Housing in general is obviously a complex topic as it deals with so many variables. If we can make a few assumptions about an average homesteader, it will be possible to nail down some specifics and thus make it easier for one inexperienced with swine to make some decisions about housing facilities.

The first assumption is that the homesteader will have from one to three feeder pigs. This greatly simplifies matters because we can talk about a growing-finishing system and forget about gestating houses, farrowing houses, nurseries, and separate growing and finishing houses.

The second assumption is that the homesteader has more time than money—but not a great deal more. He doesn't want anything super-fancy in the manner of those who substitute capital for labor, and yet he doesn't want to work any harder than necessary either.

The third assumption is that pigs will be fed out only during the most temperate time of the year. In view of the problems climatic extremes impose upon building requirements, this greatly simplifies things.

Because of the small number of animals, the desire or need to save cash, and the temporary nature of the enterprise, the most logical solution to homestead hog housing is to consider presently existing buildings. (This is also often true of larger operations where the loss of efficiency might be offset by cash savings.) To put this in balance-sheet terms, the total value of the pork produced might only be a few dollars more than the cost of the weaned pig and the feed it consumes, and if more than a few dollars is spent on housing then the pork is being produced at a loss. It would be cheaper, and certainly easier, to buy it at the supermarket.

But there might be other considerations. Spending some money on housing might make the pigs more comfortable and healthier or the caretaker's work easier. If the facilities are to be used for a number of years, the initial cost should be written off over the period of use, and the housing cost per pound of pork produced will naturally be less.

So, take a look at your existing buildings. Since the pigs will reach market weight in a growing-finishing building, the space requirements are for the larger, finishing hogs rather than the nursery-sized pigs you'll bring home. In a total confinement system this is listed as eight to ten square feet per pig. With partial confinement the shelter space should allow six square feet, and the paved outside lot should allow another six square feet. On pasture, allow twenty to thirty-five hogs per acre.

Pasturing

Pasture might have advantages for certain homesteaders. It reduces the cash outlay for protein supplements and has an obvious attraction for anyone interested in organic pork because of the natural elements grazing, rooting hogs will pick up on pasture. Housing can be extremely simple and inexpensive: in late spring, summer, and early fall, a roof for shade might be all that's needed. There is no labor involved in manure disposal because the pigs do it for you.

There are disadvantages too. Most homesteaders are land-poor, and pasturing any animal is not the most productive use for good land. Grazing animals destroy more vegetation than they consume, especially on productive land and on relatively small lots. Depending on the time and labor situation, most homesteaders would be better off to grow a crop on the land and carry it to the pigs.

Fencing would be an added expense, and it could be a headache for the homesteader with no experience in fence building. Pastures will have to be rotated to allow regrowth. Pigs will have to be trained to respect electric fencing.

While there is no labor involved in manure disposal, it probably won't be where the homesteader would like to have it, and its value will be diminished by its lying on the surface. For the ardent composter this is a serious disadvantage.

On a commercial basis two other disadvantages of pasturing swine are the slower rate of growth and the problem of parasites. Homesteaders aren't particularly concerned with rate of growth, especially if slower growth costs less money, and parasites are less of a problem on land that will only have hogs for a few months of the year.

Total confinement—the other end of the scale—doesn't fit in on the homestead because of the investment in facilities; because that system depends on the economies of scale for its efficiency where the one- or two-hog operation won't realize any savings in labor or cash; and because it just isn't natural or organic. That leaves partial confinement. Partial confinement makes sense for the homestead from the standpoint of labor efficiency, and it can also be used with many existing structures.

Partial confinement

If your homestead has a hog pen, chances are it hasn't been used for many years. Chances are also good that with a little work it can be made useable again. Even if it doesn't look like much, evaluate it very carefully—and then work up an estimate of what a *new* building of any size would cost. You'll probably decide the old one isn't so bad after all.

Of course, if the roof is caved in or the walls are rotting away, it might be more expensive to repair an old building than to construct something new. Farmers and homesteaders must be jacks-of-all-trades, and at this point having some knowledge of money management and construction is as useful as having knowledge about hogs.

Lacking a hog pen, examine other outbuildings that might be on your place. Since they were not designed for swine you will have to pay closer attention to construction, size, location, and other factors, but if they are suitable or can be made suitable with minor alterations, you'll save a lot of time and money by using them. The list might include poultry houses, dairy barns, horse barns, tobacco sheds, machine sheds, corn cribs, garages, and just about anything else with walls and a roof.

Construction ideas

And if your homestead is entirely devoid of buildings; or the buildings are already full of chickens, rabbits, and goats; or if the buildings are all falling down, don't despair: there are still ways to house your porkers without getting another mortgage on the place. You'll have to build something, but it needn't be patterned after the Taj Mahal.

If this is a first-time pig project, don't spend a lot of time building a pig palace. You might find that you and pigs are incompatible and you won't raise any more, which means the entire cost of the structure will be charged against one hog. Or, you might learn something by actually taking care of a pig and next year be in a much better position to intelligently design an efficient and attractive facility. You might even decide you want to raise a few extra to sell, and will therefore erect a larger building.

For now, think in terms of two or three pigs at the most and a shelter of about thirty-six square feet or smaller. Provide just enough room for sleeping; the pigs won't dirty their beds and the shelter will seldom have to be cleaned. If you don't have to get in there with a pitch fork, you don't need a very high ceiling, and that means you save on building materials and time (which translates into cash). Four by eight is probably about the smallest workable size even for one pig, and that will hold two quite comfortably, and even three in a pinch.

Next, look around to see what kinds of construction materials are available. Use your imagination! Don't confine yourself to thinking in terms of 2-by-4s and other contemporary building materials. You're building a pig pen, and you want to save money.

Fence posts are usually expensive, but I once bought a whole load of wooden ones for a ridiculously low price. I made a wall out of them by stacking them up. To avoid notching and other detail work that would be involved in a log cabin, I simply drove stakes on both sides at either end to hold them in place. This pen was about six by twelve, and three feet high. At one end I laid poles and boards (which most homesteads seem to have around) across the top. That was covered with a surplus sheet of plastic, and topped off with a generous thatching of "straw" left by the county crew mowing alongside the road. I did block off part of the entrance with a wooden skid from my print shop because we used this house in winter too, but that wouldn't have been necessary during the summer.

Hog houses have also been made from skids or pallets. They're available at many trucking companies for free or a small charge, or you might check out companies that receive freight by truck, such as printing plants.

I've seen temporary hog shelters made with snow fencing. No doubt field fencing would work as well. The pen was made like the log one described above but with steel fence posts placed fairly close together. The fencing was securely fastened to the posts, especially at the bottom to prevent the hog from rooting it up. Then baled straw was packed tightly around the outside of the shelter portion of the pen. More fencing and stakes were added around that to add rigidity and to keep any big bad wolf from blowing down this little piggy's straw house. Since odd lengths of fencing can often be had for a song at farm sales (and you might even have some rusted, twisted, useless-for-anything-else woven wire along one of your fence rows now if you look for it), and since the straw can later be used for mulch, this is a pretty material-efficient shelter.

Look around your place of employment, construction sites, dumps (if these paradises of recyclers haven't all been replaced by so-called sanitary landfills in your area), and anywhere else your searches take you. You'll find potential pig pens all over.

If all else fails, you'll be forced to fall back on ordinary everyday, used lumber. But even then you can be creative. Use regular frame construction choosing from a variety of designs: A-frame, shed, gambrel, or whatever meets your fancy. Or experiment with pole construction by sinking four wooden fence posts at the corners of a shed and nailing a frame and walls to that. (Tip: the pigs will be pushing *out* on the walls. Frame them on the *inside* of the poles.)

It shouldn't be hard to find thirty or forty square feet of roofing material to recycle, but if you can't find roofing or tar paper for free, check out a print shop or newspaper that prints offset and pick up some aluminum printing plates. Those from larger presses are thicker, and being bigger are easier to work with.

Please don't think I'm advocating hog shantytowns. Pigs have a bad enough reputation ingrained by centuries of tradition, and the last thing they need is a countryside full of tar paper shacks. Furthermore, I am a firm believer in the concept of homesteading, and nothing can do more damage to making this alternative lifestyle respectable than homesteaders with shoddy, decrepit, disgusting facilities. And perhaps most importantly, homesteaders follow their way of life because of a keen desire for peace and beauty, and living and working in a ramshackle environment can destroy the very essence of modern country living.

Just because something is made from recycled materials with somewhat unconventional methods doesn't mean it can't be aesthetically pleasing. This is important to the pigs,

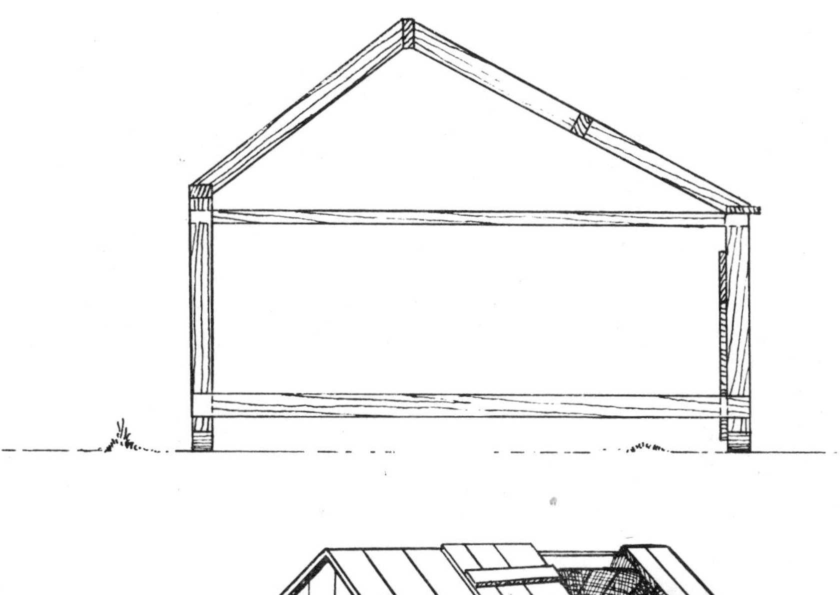

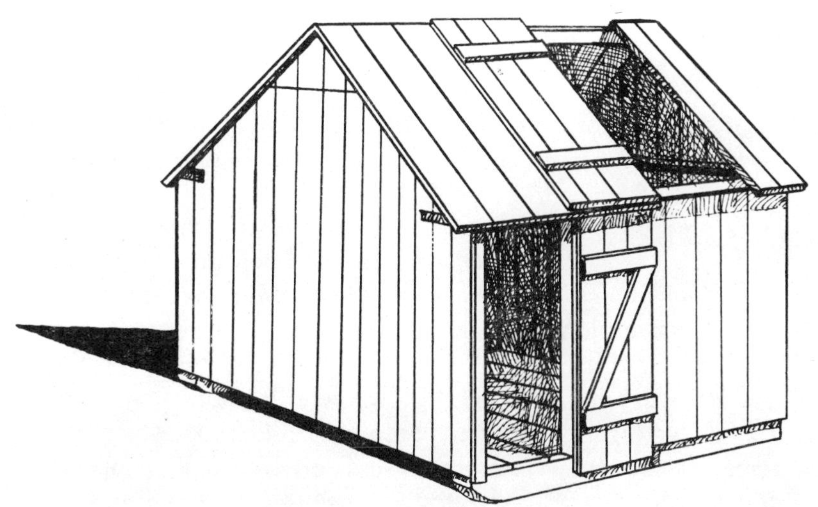

Combination-roof hog house.

the homesteader, and to all who pass by.

If you have neighbors, if you live in an area that is more suburban than rural, if you can't find useable materials, if you'd rather just spend the money to have something nice,

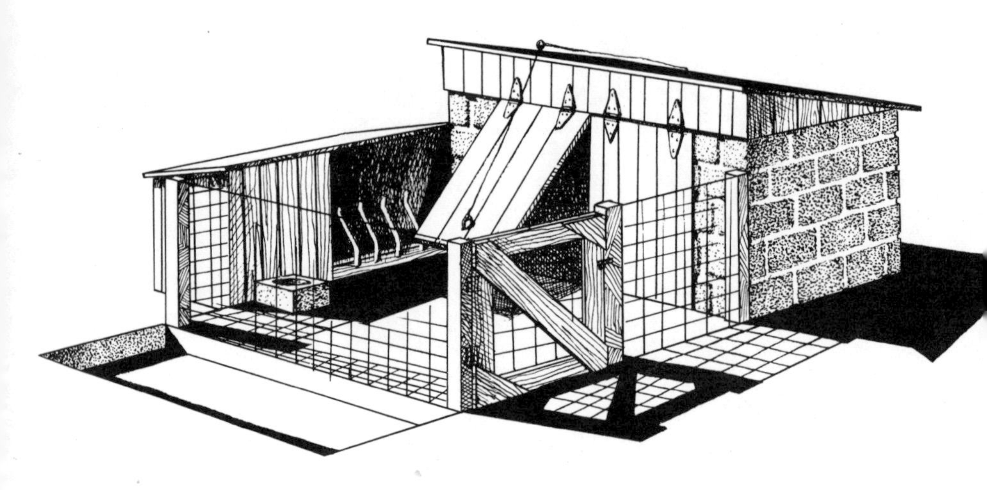

A simple hog house.

or if you're sure pigs will be a permanent part of your homestead, by all means consider a proper building.

It isn't often that I can call something "ideal," then look back on it a few years later and not make any changes. But that's just what happened with the hog pen pictured here. This five-by-eight-foot cement block building with shed-type construction and a hinged wooden front, situated on an eight-by-thirteen-foot concrete slab, just has to be the goal of every homesteader. It can be extremely attractive, and a few shrubs or other plantings will enhance the pen's appearance. It's functional, and efficient. If properly constructed it will last a long, long time, yet the cost of construction shouldn't be excessive. All in all, people who intend to raise their own food for the foreseeable future, who want to enhance the aesthetic appeal of their property now and its monetary value in the future, would do well to give this plan or some variation of it serious consideration. I would suggest an automatic drinking cup with the water pipe buried underground (and with an electric heater in northern climates). With that, and a self-feeder to be filled once a week or so, homesteaders can produce delectable pork for little cash and very little time or effort.

And that's what it's all about.

Chapter 6.
FEEDING

Swine feeding on the small place can be approached from two different avenues.

First is the "pigs is pigs" attitude most homesteaders run into when they seek help from a county agent, feed dealer, or specialist versed in any phase of modern swine production. There is only one way to feed pigs, according to this approach, whether you feed one or a thousand. And that is with ground concentrates fortified with vitamins, minerals, and antibiotics.

The other is the homestead approach. As the editor of *Countryside* I often receive letters from readers asking, "Why do we have to feed animals all that store-bought stuff: protein supplements and all the rest. Surely the farmers of a hundred years ago didn't have all these additives and their livestock survived and produced!"

One answer is that today's livestock is pushed much harder and produces more and faster, and so better nutrition is necessary. On the other hand, modern livestock produces more and faster at least in part *because* it is pushed with high-powered feeds.

In some cases the only reason farmers of a few generations ago didn't make use of some of the methods we employ today was that the required knowledge simply wasn't available. For example, nobody even knew about vitamins until the 1920s. Nutrition certainly wasn't the science it has become today. Information that is common knowledge today was undiscovered, and many of the tools and instruments for uncovering that information didn't even exist.

Similarly, many of the feed supplements in common use today just weren't available then. It wasn't too long ago that many byproducts now used for feed were dumped into rivers because no one recognized their value. In other cases, not only did no one see the need for them, but also the technology to produce them wasn't available.

And finally, the social and economic situation has changed. The world is moving at a faster pace today and life down on the farm isn't as simple as it used to be; the farm

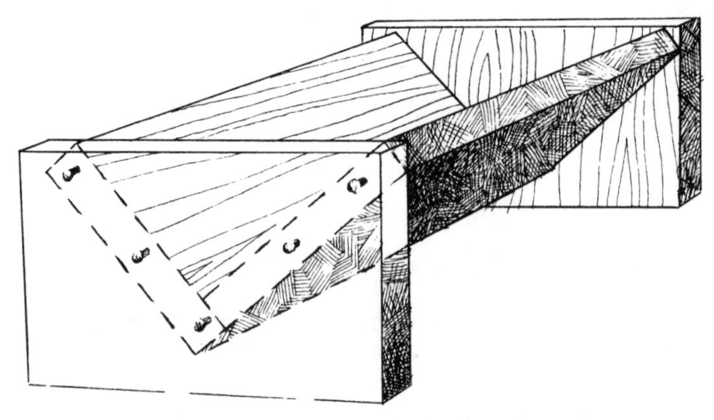

Simple trough made from two planks (2" x 12", any length) with boards nailed over the ends. Use for feed or water.

is very much a part of the world of commerce. When progressive farmers and researchers began developing hogs that had better gains and quickly reached market weights with more bloom, old-timers asked, "What's time to a hog?" Those kinds of farmers didn't survive, as farmers. Agriculture became extremely competitive, with a corresponding increase in investment, knowledge, and skill, and all-around use of technology.

Most people, even homesteaders, won't take issue with the good that increased knowledge of nutrition has accomplished. Not many people will argue with the idea of feeding wastes (such as bran) and byproducts to livestock rather than polluting the rivers with them. But ah, that third point!

Most people will say that's exactly why they're reading this book: the economic situation *has* changed; the world *is* moving at a faster pace today; agriculture *has* become big business. They don't care much for any of it, which is why they're interested in raising swine on a small scale. The agribusiness experts say it can't be done. Homesteaders know better, but they want some assurances and some pointers. Here are a few.

We must bear in mind first of all that penned animals depend on their caretakers for a balanced diet. Swine, especially in confinement, have less choice in their feed than any other class of livestock.

Moreover, good nutrition is extremely important for swine because they grow at a much faster rate than other farm animals; they also mature and produce young at an earlier age, and produce *more* young. Yes, in a way they're forced. But the simple truth is, without these attributes pork would be even more expensive than it has been recently, and few people, homesteaders included, would be able to enjoy it and benefit from it.

However, this doesn't mean that homesteaders can't combine their ideals with the facts of life, make judicious use of applicable technology, and come out ahead. There

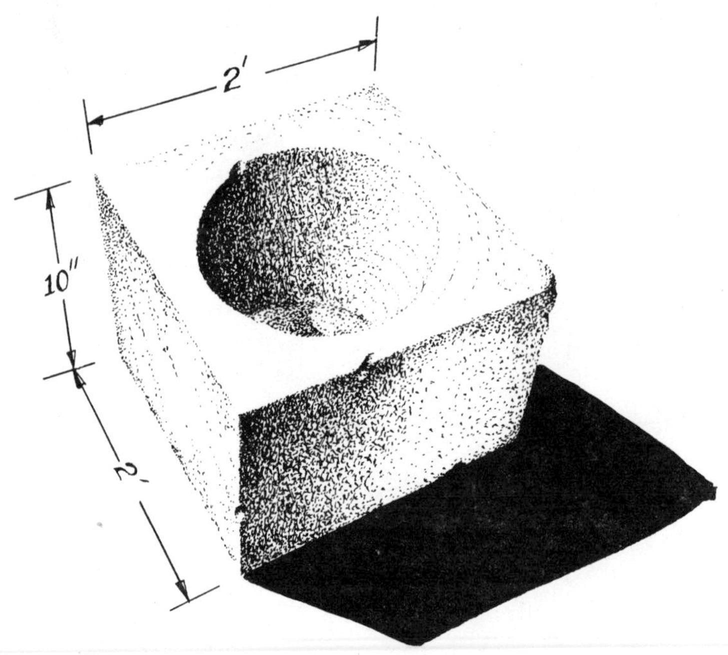

Pour concrete 8"-10" deep into a wooden mold about 2' square. Depress a 5-gallon can in the center. Hogs can't move or overturn this feeder-waterer.

are two sides to this: the socioeconomic, and the nutritive. First we must understand the forces acting on commercial hog production, and how they differ from the management goals of the homesteader.

How commercial production varies from the homestead operation

We've mentioned that commercial farming is highly competitive, and that farmers have understood they must make use of all the technology possible just to remain in business. The hog farmer's competition comes from other hog farmers: the one who produces at the lowest cost obviously earns the highest return. But it also comes from many other areas. If the price of pork in general gets too high, consumers will choose chicken or beef, or macaroni. The farmer's profit can also be affected by government policies, imports, and production costs, all of which put the squeeze on the guy who is trying to make a decent living doing an honest job.

Under these conditions the alert and capable farmer-businessman will naturally make use of all the aids available to him; in general, this means all the research results that show how to produce the most pork at the least cost. These include advanced breeding techniques, concentrated feeds and feed supplements, confinement housing, hedging on the commodity market, and many others.

However, as we've noted earlier, many of these management devices give rise to other situations which require still more technology. Some examples would include the fact that hogs in confinement require feed supplements that might not be needed if they were allowed free range; that breeding for meat type might give rise to problems with pale, soft, watery pork, which requires still more technology; and that confinement housing invites tail biting so tails are docked and then the pigs bite each other's ears.

The point is that while technology itself is neither good nor bad, it must be used with discretion, and what's more, there may come a time when the expense of technology might bring about results just the opposite of those the technology was designed for!

This is what the homesteader is looking at, and there is some interesting data to support his position. But before we get into that, let's look at how the homesteader's methods and goals differ from those of the commercial farmer, especially as regards feeding. To avoid comparing apples and oranges, let's assume that the homesteader considers the cost of buildings and labor just as the commercial farmer does, and uses commercial feeds. As most agribusiness experts are quick to point out, the cost of feed alone is enough to make the one-hog enterprise unprofitable in most cases. Grains will generally cost more if they are purchased and not homegrown: grinding and mixing costs for small lots are higher per pound than they are for large batches. There's no law that says the homesteader must use commercial feeds, but it makes this comparison more accurate.

The homesteader will buy a feeder pig at market price or perhaps even above, again because of discounts on quantity. The housing and equipment costs per pig will be higher in some cases if only because the facilities won't be used as intensively. Hours spent per pound of pork produced will be higher because of the economies of scale but also because the homestead is not as automated.

All of this would make it appear that the homesteader is at a tremendous disadvantage in comparison with the commercial feeder. And so he is—if he tries to raise pigs the way the commercial feeder raises pigs.

For some homesteaders this disadvantage doesn't exist. They liken the commercial producer who runs a pork factory to any other manufacturing concern, say a company that manufactures cheap furniture. The homesteader's hog is more akin to a piece of handcrafted furniture. With the right materials and the proper skill and experience, the handcrafted chair will be attractive, sturdy, and will embody a great deal of pride and satisfaction. It costs more because it's worth more.

But things don't necessarily work out this way, and here is where we can begin to see some of the effects of technology.

The commercial farmer invariably has to borrow money to finance his operation. In Illinois more than 50¢ per bushel of corn produced goes to pay interest, and that's just one debt involved in producing the corn. Many large farmers have hefty interest charges on their buildings, equipment, and even stock.

The homesteader, working with a much simpler system and much smaller figures, will pay cash.

The farmer profits or loses on a hundred, a thousand, or ten thousand head. A dollar-a-head loss on ten thousand animals might mean catastrophe: a dollar-a-head on one animal means a couple of packs of cigarettes. Without a profit, a farmer can't run his business and he can't buy shoes, food, and clothing for his family. The homesteader can feed a pig with no return on labor and not even notice, not only because the check from town still goes to the bank every payday but because the few minutes it takes to care for the pig are as insignificant and as enjoyable as walking the dog is for some urbanites.

The farmer faces a greater risk of losing animals and incurring expenses related to health. He deals in a larger universe which increases his chances of loss on a statistical basis; the homesteader who chooses one pig is more likely to get a healthy one than the fellow who chooses a carload lot. Also, congestion and sheer numbers make the commercial pigs more prone to disease and injury.

So in a detailed accounting it might well be that the homesteader actually comes out ahead. We could carry this further to show again that bigness requires still more growth and technology demands still more technology. The homesteader is concerned with using what knowledge and technology proves to be useful in terms of his goals and methods, and ignoring the rest. To intelligently develop such a system we must go back to the point where large-scale enterprise veered sharply away from old-fashioned homestead methods, then add in whatever new developments fit the modern homestead situation.

Planning a diet for the homestead hog

Homesteaders will have few arguments with the genetics of the modern hog, and they couldn't do much about it even if they did. We're still talking about buying one or two pigs from a breeder who produces for a market that is much broader than the homestead market.

Homestead hog housing is considerably different from the ideal commercial housing, and this has many implications. The first, already mentioned, is cost. There is less investment and therefore less interest, less depreciation, lower taxes, and lower operating and maintenance costs. Because there is less expense in housing, and because of homestead goals and methods, homestead hog facilities can logically be designed differently from commercial housing, allowing more room which will tend to help eliminate certain stress and health problems. This in turn can permit a slightly different type of diet to which we can add the different rations that are available, useful, or desirable on the homestead anyway: garden waste, kitchen waste, dairy waste, and surplus of all kinds.

At this point we have a situation already far from the modern commercial operation, and the differences don't stop here. Because of the difference in the labor structure, the homesteader can afford to provide nutriments that would be uneconomical for the large operator. In addition to milk which has proven food value, these might include comfrey, the only land plant containing vitamin B_{12}, which is an element of great importance to baby pigs, and pumpkins, which are high in vitamin A. We'll delve into this in more detail when we examine feeds more closely, but from this brief account it can be seen that just as technology demands more technology, natural methods permit the ever-increasing use of natural methods. The homestead hog raiser can circumvent the marketplace for money, energy, and fossil fuels; a great deal of feed or in some cases all of it; medications and other chemicals; and can even do without the marketplace for marketing, which includes everything from transportation and slaughtering to supermarket advertising and checkout clerks.

In fact, even the commercial producer might to some degree profit from examining the merry-go-round that technology has put hog production on. There is a "which comes first, the chicken or the egg?" atmosphere about both high technology and organic systems, but as the use of technology becomes more questionable and more expensive, perhaps it's time to step back and assess the current status and see what might have gone wrong.

Yes, indeed, the homestead hog raiser can compete.

Old-time methods

I have a delightful old book, written in a scholarly fashion (for our day, at least), that is 1,400 pages long and has a title to match both its tone and length: *Livestock: a cyclopedia for the farmer and stock owner including the breeding, care, feeding and management of horses, cattle, swine, sheep and poultry, with a special department on dairying, being also a complete stock doctor,* by A. H. Baker, M.D., V.S. It was copyrighted in 1909. I treasure it for many reasons, not the least of which is the way it enables me to put modern methods in some perspective. It also obviously is a rich source of information on what we today would call homestead methods.

There are many revealing passages in a book that's almost seventy years old, but here is one of the most poignant dealing with organic homestead hog raising today:

"I believe that the purely artificial breeding and feeding of breeding stock, the indiscriminate ringing, the absence of roots, and the feeding of breeding animals almost exclusively on corn, have, in many cases, so enfeebled the constitution of swine that they have become an easy prey to the various epidemics and contagious diseases that, of late years, have carried off so many. And I believe, also, that the utmost care will be necessary in the future to guard against this disability."

The practices the good doctor objected to even in 1909, and in particular those he advocated, should be of great interest to the organic homesteader of today. What's even more encouraging, we will see shortly that many of his ideas fit in with those held by later authorities.

Dr. Baker mentions that, even then, there were only rare cases where swine had unlimited forest range. Lacking that, he maintains that the greatest profit "lies in forcing their feeding to fatness." He also points out that "true economy will dictate that they have the warmest possible shelter in winter, and that they be kept cool in summer." So there are a couple of items that haven't changed. As we've indicated before, though, warm winter shelters and cooling in summer have been carried to extremes that were certainly unforeseen in 1909, and have brought with them other problems.

Be that as it may, Dr. Baker also lays great stress on certain conditions which he believes important and yet neglected, and which are totally overlooked today in the name of science and economy. Here is a synopsis of what he calls the proper food for swine:

He notes that swine eat fewer varieties of herbs and grasses than any other animal, although they will consume pigweed (amaranth), red and white clover, and a few succulent plants. The natural foods of the hog, Dr. Baker tells us, are artichokes and other bulbous roots, many insects and grubs, frogs and such small animals as they can kill, all the grains, and culinary vegetables. "Such then, is their proper food—so far as it may be obtained—if the highest constitutional vigor is to be preserved in the breeding stock. If they are allowed a fair range on clover, including the gleaning of grain fields in summer; and if a good supply of pumpkins, and the refuse fruit of the farm be allowed them in autumn; and if in winter they be allowed daily rations of artichokes, small potatoes, parsnips or carrots, they may have, in addition, what grain they need to keep them in full flesh—not fat. If this course of feeding were generally adopted for the breeding stock, we should in a few years hear but little of the epidemics which periodically sweep the swine away by thousands. But as long as there are so many breeders who never look beyond present profits, these epidemics will probably continue to be bred among the herds of this class to scatter the germs far and wide." (I can just hear the organic farmers and homesteaders of today chorus, "Right on, Doc!")

He also advocates giving young pigs all the milk and slops of the house. Other feeds he recommends are alfalfa (although that crop was just becoming known in this country then), field peas cut just before they shell, rutabagas, beets, and mast (which is acorns and similar forest-floor fare). In the fall, he says, there is no better food than pumpkins and grain boiled together. Grain is the cheapest food and the main dependence of the farmer, but the others are useful for keeping the animal in good health and digestion, for without good health and sound digestion, no animal can be made fully fat. (His choice of words.) Remember that this predates vitamins and other concerns of nutritionists. Yet his approach almost seems to embody an instinctive regard for unknown qualities. (Note that there are unknown qualities today—it's significant to realize that we still don't know everything about nutrition.)

In another pertinent passage the author admonishes every farmer to raise and cure enough hogs to furnish pork, bacon, and hams for home use. "It will be found, in nearly every instance, cheaper than to buy the bacon and pork already prepared."

He also suggests that family-size pork raising be in "close pens"—what we now call partial confinement. He gives a pen size of twelve by sixteen feet as adequate for six large hogs, divided into a "sleeping room" and a "feed room" both eight by twelve.

As for health and sanitation, he describes a barn in which he once raised five hundred hogs, "with but slight loss from epidemics." He points out the importance of an ample supply of pure water, then continues, "The pens were kept regularly washed; the offal was carried to the compost heap and covered regularly with earth; and the hogs had always by them ashes and salt, and also a supply of bituminous coal slack."

The value of grinding grain for hogs was recognized by 1909. Dr. Baker cites feeding experiments which showed that one pound of gain in live weight required 1.2 pounds more of whole wheat and oats than was required with ground grain under the same feeding conditions. As much as 50 percent of the whole wheat went unmasticated in those early feeding studies.

Other authors recognize the value of grinding grain but also concede that many farmers had no means then of grinding it, a situation very prevalent among modern-day homesteaders. In that case the grain should be soaked for twelve to twenty-four hours. Baker acknowledges that the loss is less if the grain is soaked, but still maintains that it's a poor second to grinding.

Cooking grain for hogs, at one time a common practice, is considered of "doubtful economic value" by 1909. Vegetables are a different matter though, according to this writer. He claims the feeding value as well as the palatability of most vegetables is improved by cooking. Potatoes are given as one example, citing tests where hogs were fed 12.4 pounds of potatoes mixed with 2.8 pounds of chopped oats and shorts mixed half and half. (Shorts, or middlings, are a byproduct of wheat milling that consists mostly of fine particles of bran and germ.)

Baker is also high on pumpkins for hog feed. In one experiment (at a government station in Oregon), pumpkins were fed with shorts at a ratio of about one pound of shorts to eight pounds of pumpkins. At that time, it was the cheapest pork produced from any combination of food materials tested at that station; the pork was pronounced first class for bacon purposes; and the hogs were healthy.

Baker was located in the Pacific Northwest, and he maintains that, "Our methods of feeding, together with a greater variety of food material, is conducive to the health of the animals. The comparatively small proportion of corn fed is an advantage to health. Corn, being a highly carbonaceous food, induces more animal heat, and should be mixed with some food rich in protein to give the best results, both as affecting the health of the animals and the quality of the product. Where wheat, oats, barley, peas, alfalfa and clover constitute the chief food supply there is little danger of disease. . . . With good blood in the herd as the first essential, and then a proper food supply, the results will be wholly satisfactory."

This is just a brief recap of some of Dr. Baker's ideas on hog raising around the turn of the century. Other concepts of his, and of other early writers on the topic, are woven into various sections of this book and serve several functions. First, they provide specific information as well as general ideas for those homesteaders who honestly want to go back to the older ways of doing things. And secondly, they provide a starting point for those who instinctively feel there is something wrong with certain present-day assumptions and methods, and yet are reluctant to toss out the baby with the bath water. In order to save what is good in the present system they must be able to *recognize* what is good, and why the undesirable elements are undesirable.

Incorporating the old with the new

I find these old-timey ideas as valuable as they are fascinating. How did we get here from there, and why? The role of social and economic factors very often appears to be greater than the role of technology so far as impact on agriculture is concerned, and that should be of significant importance to homesteaders. By raising a pig or two in our backyards we aren't really bucking modern agriculture per se as much as we're saying, "I don't like the socioeconomic factors that decree what kinds of food I eat and how that food is produced."

We've already mentioned some of these factors as they pertain to swine. The competitive pressures on farmers today are considerably greater than they were in 1900. We've seen how hogs have changed in response to consumer demands. It seems logical that technological methods have come about because of these forces.

It's not too hard to anticipate some of the objections modern farmers and other agribusiness people would have with the hog-raising methods of the early 1900s. Even homesteaders might have a couple, based on changing conditions. There is evidence that things were changing already in 1909.

One of the most noticeable was the slaughter weight of hogs. Baker mentions farmers feeding hogs to six hundred pounds—which required three years! But he also points out that it's altogether cheaper to feed three hogs to two hundred pounds than one to six hundred pounds.

Some other points might be subject to a great deal of debate. For example, those two-hundred-pound animals in 1909 were nine months old, while today they wouldn't be more than five or six months old. Early gains are the cheapest, a fact that was common knowledge even in 1909. In other words, the more slowly the animal grows, the more feed it will take to reach a given weight.

Commercial producers are adamant about the value of pushing the pigs because of the investment they have in facilities (they must market as many pigs as possible per year in order to amortize building and equipment expense on as many animals as possible), and because of the structure of their feed costs. The first doesn't concern many homesteaders. The second point bears closer examination.

Hog farmers of 1900 did not feed soybean oil meal (at $200 a ton), antibiotics, and vitamin and mineral mixtures—and it took the hogs almost twice as long to reach market weight as it does today using these feed supplements. Modern farmers consider this time lag important because of feed costs: although the supplements are expensive, they save more than enough grain to pay for themselves and have the added benefit of keeping the hogs moving to market at a faster clip.

The homesteaders who are most concerned about the use of high-powered feeds may or may not have the same feed price structure. If they buy grain, they're in an even tighter price squeeze than farmers who either produce their own grain or buy in volume. But if homesteaders produce their own grain, and can do it economically, they might be ahead. This obviously depends on a great number of factors, but to take an extreme example, say a homesteader plants and harvests an acre of corn—by hand. His cash costs are practically nil. (The county agents are snickering at this but lots of people do it, and it could make economic sense for many more.)

Then carry it a step further. Let's imagine a homesteader who ponders Dr. Baker's advice and decides *not* to feed his pig in the conventional manner. He grows a little corn, some pumpkins, some Jerusalem artichokes and potatoes, a patch of alfalfa or clover, and perhaps a few other goodies just for the pig. In addition the hog gets garden and kitchen surplus and waste. Now what happens?

The animal's requirements for feed supplements and additives decrease. The homesteader's cash investment in feed crops goes way down. The cost of the pork plummets—and it's organic!

The fly in the ointment is labor, and that's one of those socioeconomic things that is often difficult to pin down. The above system won't work with large numbers of hogs because it doesn't lend itself to automation, and without automation the man-hours per hog are too great to be competitive. But on the other hand, is it possible that it might enable a homestead farmer to make dollars per head on a few animals, rather than the pennies per head involved in large-scale production? The choice between high-tech and low-tech management systems is clear, for very large operations and for very small ones. But is it possible that with proper methods and proper accounting, the low-technology methods might prove profitable on a much larger scale than now believed possible?

This question certainly deserves attention. Today most of the people (in this country at least) who are involved in low-tech agriculture are doing it for "fun." Planting a quarter acre or so of corn by hand can be an enjoyable after-work or Saturday afternoon activity. Harvesting that corn and feeding it to a couple of hogs is no real job for people who push buttons or shuffle papers or have any type of confining, indoor, mental occupation. And this can be expanded.

First, there is a growing awareness that many people filling those indoor mental jobs would really rather be planting corn and feeding hogs! There are also the unemployed, who we seem to take for granted as a necessary evil of our economic system today: does it make sense to support them from public funds when they could be increasing the cost-efficiency of world food production, and garner satisfaction and self-esteem in the bargain? While hog production, like everything else, will always be based on economic criteria, there might well be social factors that should be considered.

And there are certainly economic changes that make a low-technology operation look more attractive. The cost of money is of great importance in large operations, and not only have interest rates been at high levels, investment in agricultural enterprises has progressed at a budget-busting pace. Scaling down has obvious implications, but because "efficiency" has always been related to scale, few people pause to ponder those implications.

Then there are fertilizer costs and all the other expenses connected with large-scale, high-technology farming. Some people don't know that most chemical fertilizers are made from petroleum, but once they find out, they get the picture in a hurry. Smaller, organic operations are less dependent on fossil fuels and nonrenewable natural resources of all kinds, so as oil and other materials become more scarce and costly, the economic balance shifts in their favor.

These are just some of the considerations that make it exciting to examine hog raising in 1900, hog raising in the 1970s, and then through logical analysis chart a course for all-around sensible hog raising in the future. Our primary goal as homestead hog farmers is to raise a couple of swine in the best way possible. But we might also be laying the groundwork for something much more important, and lasting.

The foregoing gives us something to think about while

we're shoveling manure, but it also gives us the assurances we need: it *is* possible to raise a pig or two and come out ahead.

Figuring the cost of homestead-raised pork

How much feed will it take to produce the pork that will show up on your homestead dining table as succulent ham, crisp bacon, and mouth-watering chops and roasts? Researchers at Iowa State University say that each market hog has accounted for 891.1 pounds of feed, broken down as shown in the table:

Pregestation and gestation ration	133.6 pounds
Boar ration	5.5
Lactation ration	49.2
Starter ration (to 25 lb.)	14.2
Grower ration (to 50 lb.)	52.5
Developer ration	248.8
Finisher ration (to 220 lb.)	387.3
Total	891.1 pounds

These figures are based on the following assumptions: 1.8 litters per year; 13.2 pigs per sow marketed per year; one boar per forty sows; 4 percent mortality after weaning; and a loss of feed in processing from 2 to 15 percent. Note that these figures include feed for the 20 percent of the gilts kept that don't breed and also take into account sows that conceive but don't farrow, sow mortality, lost litters, and feeding gilts that will go into the breeding herd. They demonstrate rather graphically the importance of management: the most room for improvement in converting feed into pork is in the area of getting bigger, healthier, more regular litters from sows.

The one-hog homesteader isn't overly concerned with most of this data. Sow management is up to the farmer who farrows pigs and sells the homesteader a weaner. The homesteader has no control over it. What's more, considering the way agricultural marketing works in this country, the management ability of the farrowing operator won't have any effect on the price the homesteader pays for the weaned pig. The pig will probably be sold at current market levels, with no regard for the actual cost to the producing farmer. That means the farmer might make money, but he's just as likely to lose money. Management quality is important to him to assure maximum profit as often as possible, and minimum losses at other times. Naturally the homesteader who keeps sows will be in the same position.

But the one-hog homesteader will be interested in the feed consumption figures beginning with the grower ration. According to the data presented here, that amounts to 688.6 pounds. Note that this is starting with a twenty-five-pound pig; it includes provisions for 4 percent mortality which is rare on homesteads; and it also provides for waste and processing losses. On the other hand, it assumes that antibiotics will be fed with a resulting increase in feed efficiency, and of course it's based on grain. Forage and succulents that contain a much lower percentage of dry matter will raise the total weight of feed consumed. To illustrate, if very young alfalfa has 18 percent dry matter, it would take one hundred pounds of it to provide eighteen pounds of "feed." The rest is water. This is one reason grains are so important. A pig couldn't eat enough forage to get the nutrition it needs, even if the forage contained all the required nutriments, which it doesn't.

To determine what your pork will cost you on a homestead basis, you need this information: the cost of the pig; the cost of feed in your area during the feeding period; and the cost of any supplements fed. Add in housing and equipment costs if you like. Most homesteaders don't bother because their costs are so low or so difficult to specify because the facilities are there anyway. Likewise, few homesteaders bother to charge for their labor, management, or capital investment.

If you will have the pig butchered, find out the cost in advance. I've already paid more for the butchering than I paid for the pig and the feed, which definitely makes it a losing proposition.

Divide the total of all these figures by 135 pounds—the amount of meat you can reasonably expect to get from a 220-pound hog. That will be the cost of your pork.

Don't forget that a pork carcass is not all ham and pork chops! You'll probably come up with a figure that will look pretty good when compared with the cost of bacon in the supermarket, but the carcass also contains lard, trim, and less expensive cuts. You can figure on about 23 pounds of shoulder, 27 pounds of loin, 31 pounds of side (of which 24 pounds are bacon, 2 pounds are sausage trimmings, and 5 pounds are spareribs); 31 pounds of ham; and 36 pounds of miscellaneous, which include feet, tail, neck bone, jowl, and fat. There will be about 120 pounds of actual meat that can be considered retail cuts.

The actual profitability of home pork production depends on many factors, but the cost of feed is paramount. The exclusive use of commercially prepared feeds makes the profit outcome doubtful, but the only way to actually determine how much use you can make of homegrown feeds—extra milk, vegetables, eggs, etc.—is to raise a pig and keep records on costs. With a little knowledge you can raise good pork on the homestead, and cheaply.

Chapter 7.

PROVIDING AN ORGANIC DIET

Anyone who wants to raise swine organically isn't likely to get much help or encouragement from the traditional sources of information: county agents, universities, and feed companies. And to some extent, anyone raising swine on a small scale with *any* method faces the same situation. Maybe the agribusiness folks won't openly laugh at you, although that's been known to happen. Even if they don't show open hostility, about the best you can hope for is tolerance and, in a few cases where one of them is sincerely trying to be helpful, a certain amount of confusion.

The hog business has traditionally been a conglomeration of small producers, and in a way it still is. But farms marketing a thousand head and more per year are becoming more common, and some experts predict that ten-thousand-head production units will be the norm before long. Because of the tremendous importance attached to rapid gains, feeding, housing, and saving labor, and because all of these have more application to large-scale operations, anyone who isn't big and getting bigger is sort of left out in the cold.

We can't really blame agribusiness for reaching its present state. Considering its goals—providing more food more efficiently, especially in terms of human labor—it has done an admirable job. Unfortunately, however, it doesn't address certain other factors such as chemical residues and other toxic materials in food; the human side of agriculture and life in general; the accounting system that allows the world to live off its capital such as soil and fossil fuels and think it is creating wealth; and similar concerns of the organic farming segment of our society. Not only does agribusiness ignore these factors: it's in an active struggle against them because admitting their importance would make it more difficult to reach its own primary goals.

Ironically, while the goal of agribusiness is to produce more food more efficiently, its single-minded dedication to that goal causes it to ignore other concerns—the concerns expressed by organic farmers. And the main tenet of organic farming is that agribusiness is on a collision course with calamity, which means *less* food will be produced and *less* efficiently!

In other words, the basic goals are the same for both groups. Dr. Baker aptly describes the agribusiness group of today even though there were no chemical fertilizers in his time: they "never look beyond present profits." Apparently there have always been farmers who haven't taken their stewardship seriously, who have failed to grasp the implications of their work for the future of our species.

So far as swine are concerned, it's possible to adjust our thinking in terms of the ultimate goal and still take into account the possibly deleterious side effects of bulldozing our way to that goal. There are some definite signposts marking the path that perhaps *should* have been taken between the publication of Dr. Baker's book and the latest issue of any farm magazine that's packed with ads for chemicals and $20,000 tractors. There are facts that support the contention that we've moved too far too fast—and in the wrong direction. A knowledge of those facts is essential to those who want to get back on the right track.

Pasture vs. confinement feeding

Pasture is of considerable interest in this regard. In 1909 Dr. Baker placed great importance on the role of pasture in swine nutrition in spite of his realization that swine consume little such roughage. The role of vitamins was unknown when Baker wrote his book, and if he had known of their importance he might have laid even more stress on forages. He didn't know about vitamins, but he knew that hogs did better when they ate clover and alfalfa because he could see the results.

That intuition was backed up by Frank B. Morrison in *Feeds and Feeding*. First published in 1917, this book is still in print (being continually revised and rewritten) and is still cited as the undisputed authority in the field of animal nutrition.

"Why do we have to feed store-bought additives to swine?" the homesteader asks.

"There will generally be no deficiencies of vitamins when swine are fed sufficient well-cured legume hay or when they are on good pasture." He also maintains that green forage and pasture conditions are definitely superior to legume hay in meeting nutritive essentials.

And again: "Adding alfalfa hay or other legume hay to corn and minerals for pigs not on pasture will generally permit fair gains and prevent serious nutritional trouble." Gains will be less rapid and more "expensive" (with current accounting methods) than if soybean oil meal or a similar supplement is added, but the homesteader's question is answered. If you are not paying an astronomical sum in interest on your buildings and for other reasons don't have a consuming concern with moving hogs to market just as quickly as biologically possible, there is no need to spend cash on commercial additives.

While feeding hay (or better yet, fresh-cut green forage) is helpful, it isn't the complete answer. Morrison also mentions "unidentified vitamins or factors," a term commonly heard in connection with hog feeding. The technologist gets acceptable results (by his criteria) with the products of the chemist, but even the chemists must admit that there indeed are "unidentified vitamins or factors." The organic farmer supplies them by pasturing or, in partial confinement, by feeding green forages or hay, and by giving the pigs access to a supply of uncontaminated earth. It has been said that young pigs especially need these unknown factors.

The high-tech hog farmer assumes that his rations, formulated by experts in nutrition and chemistry, are adequate. Even if they do cost more than dirt, they at least allow the use of other technology and so are said to save money. But back on the organic farm, the thoughtful person is taking a new look at some old knowledge. And he's getting support from some surprising sources.

As of 1974, one of the nation's leading hog nutritionists, T. J. Cunha of the University of Florida, stated, "I realize many scientists are of the opinion that unidentified factors do not exist. I feel they do." He said plain dirt lots are better for pigs than confinement is, but that pasture is best—a contention based on observations made over an eight-year period in many different areas of the world. Even if his statement is controversial and if some sectors of agribusiness might not agree with him, he has an impressive record. He is a member of the National Research Council (NRC) which works out nutritional requirements for livestock; these are used by feed manufacturers and farmers. And he met opposition when he advocated higher levels of choline in sow rations to correct spraddle-leg conditions in baby pigs. Higher levels of choline in sow rations are now common, and there are fewer reports of spraddle-legged pigs.

Because feeder pigs go to market at about five months of age, slight nutritional deficiences are difficult to assess. But Cunha cited some interesting facts about sows. After being in total confinement for a year or two, sows produce fewer litters because of a lower conception rate. The farmer who raises sows in confinement is also experiencing more services per conception, increased sterility, poor feet and legs, and other problems, and also infantile reproductive tracts in their gilts. Gilts raised in confinement are less likely to show signs of being in heat, and have delayed puberty and reduced productivity, Cunha said. Boars don't last as long in complete confinement as those not confined, even with a "well-balanced" diet fortified with minerals, vitamins, and dehydrated alfalfa meal plus green chopped alfalfa in the summer.

He has observed that the longer sows are kept in confinement, the more problems occur. In general, he said, sows kept under tethered conditions have done the poorest of any he's seen. "Reproduction appears satisfactory when sows are first moved from pasture into confinement, but reproductive problems usually start to occur after the first or second litter." The reason for this is that animals apparently store factors which must be depleted in the sows before research can demonstrate the results of their lack in the diet, and, according to Cunha, that's why some researchers have been unable to demonstrate the importance of these factors.

Some researchers have little faith in studies where in one trial alfalfa meal, fish meal, or some other source of the unidentified factor proves beneficial to the pigs, while another trial with the same source shows no benefit. Cunha believes that is explained by a variability in the presence of unidentified factors in feed: some alfalfa meal contains them, and some does not.

While Cunha recommends a high quality diet with properly fortified protein, minerals, and vitamins, he believes pastures can cover up many omissions in feeding. As one moves away from pasture, a properly balanced diet becomes more important—but if there are indeed unidentified vitamins and factors in a balanced diet, how can a complete ration be formulated in a test tube?

The organic farmer who sees his goal as not only producing food, but *good* food, will treat this information with greater interest and respect than will the farmer who thinks everything can be explained in terms of test-tube analyses and dollars. There is a common saying that if nutrition isn't in the soil, it can't be in the food grown on that soil. Similarly, if the diet of an animal is lacking an essential element, that element is not available to the human consumer. This was demonstrated at the Washington Agricultural Experiment Station in work that showed a positive relationship between thiamine intake of a hog and the thiamine content of pork. One pork chop from a pig consuming thiamine-enriched feeds was shown to contain the minimum daily requirement of thiamine for a human. However, a pig on a low-thiamine diet produced chops low in thiamine, and it would have taken ten of them to meet a human's minimum daily requirement.

We need healthy soil to produce healthful food for livestock and humans. The livestock must obviously be healthy too, but to repeat it one more time, how can we concoct a truly balanced diet if there is good evidence that we can't even identify, much less isolate, certain factors in such a diet!

No doubt scientists will continue to solve this and similar problems in the laboratory. They will also place considerable emphasis on breeding and selecting animals that will do well in complete confinement. In spite of the fact that at the present state of the art, each new solution seems to bring with it a myriad of small but nagging problems, they'll continue their pattern for a number of reasons. Chief among these is the economic pressure exerted by that goal of "more food more efficiently." However, even that might be changing.

The well-publicized report entitled, "A Comparison of Organic and Conventional Farms in the Corn Belt," released in July 1975, showed that organic farming can produce about as much per acre, earn as good an income, while using about one-third the energy of conventional farming. As energy becomes more expensive and organic methods become more sophisticated, the balance could conceivably tip in favor of organic methods. This includes pork production. In fact, it's possible that we're very close to that point now, but few people realize it because nobody has "made a study" using the right criteria.

The organic farmer and homesteader, large and small, can read a report of some new finding or technique and evaluate it in light of their viewpoint. If enough people do this and apply the results, the day will arrive when the organic pork producer and the homestead hog farmer will not confuse and amuse the experts when they seek information. Rather, the experts will see the light themselves and redirect their efforts in the proper channels.

While this might seem like an impossible, optimistic dream right now, there is ample evidence that anyone who wants to produce organic pork need not have an inferiority complex.

Homegrown vs. commercial feed

How then, does one feed a pig?

The easiest way is to go to a feed mill. Depending upon the age and weight of the pigs, the mill will mix up a combination of grain and supplement containing 12, 15, or 18 percent protein, as well as minerals, vitamins, and antibiotics. While this might be an especially attractive course of action for the homesteader who hasn't the land, machinery, or the time to produce feed, it can also have drawbacks. Many mills will be reluctant or unable to mix very small batches of feed, and there is almost sure to be a price penalty, even if it is measured in pennies. It takes the same amount of labor and overhead to run a mill at full capacity as it does to run one with a fraction of a load, and there is likely to be more expense in measuring small amounts of ingredients. For example, a miller who customarily uses so many bags of premix per load and who has to break a bag and weigh a few pounds for you, not only takes additional time—he has a broken bag sitting there.

There is another problem with feed from a mill. The commercial feeds *are,* for all practical purposes, complete, and that means that when other homestead feeds are added, the carefully formulated balanced diet is thrown out of whack. On the homestead, other feeds are available and should be used, if only because their use is part of the economic rationale for keeping a pig.

In spite of all this, purchasing commercially prepared feeds might well be the wisest course for some homesteaders. Or, a farmer might grow corn and have that corn ground and mixed with supplement at a mill. (As a general rule of thumb, ag experts say you need to use at least seventy-five tons of feed a year to economically justify grinding your own feed with an on-farm grinder-mixer.) Or, a homesteader might use a varied diet approach based on Dr. Baker's suggestions.

Because Dr. Baker's ideas will have a great appeal for many homestead hog raisers, because we have seen that they can be substantiated at least in part by subsequent research, and because they can have great practical possibilities on many homesteads, we will recapitulate them here and then examine them in greater detail.

His basic premise is that the hog—like man—is a universal feeder. "Any common sense man may see . . . that it is true economy for every breeder to spare no pains in providing for this class of livestock diversified food which they crave, and which is necessary to make sound and vigorous constitutions." To which the organically inclined will add, "and which is necessary to produce food fit for humans."

Corn

Corn is at the heart of the matter. Normally, over one-half of the U.S. corn crop is fed to swine. "The corn-hog ratio" is a common term used to show the number of bushels of corn needed to equal the value of a hundred pounds of pork on the hoof; in other words, corn and hogs go together like love and marriage. The normal corn-hog ratio was 13.8 from 1940 to 1964, meaning that the cash value of 13.8 bushels of corn was equal to a hundred pounds of hog. A higher ratio means corn is cheap and the farmer might be better off raising hogs than selling corn. A lower ratio means the opposite. (This ratio has fallen into some disfavor in recent years, however, because of changing cost patterns in both corn growing and hog raising.)

Hog feed to many people means corn, and more than a few bulletins claim that hogs will do well on a diet of nothing but corn and supplement. Baker, however, maintains that a mixed diet will give better results than any single grain, corn included.

The popularity of corn is due to the fact that it is an excellent energy feed, high in digestible carbohydrate, low in fiber, and very palatable. But most of all, it is cheap. So we get back to the economics of modern agriculture.

Corn is cheap, or we think it so, for several reasons. Tremendous yields are possible with corn. The world record is 338 bushels from one acre, set in 1975 in Illinois, and in normal years the average exceeds 150 bushels per acre in prime corn-growing regions. Contrast that with yields of 30, 40, or 50 bushels per acre of other small grains such as oats, wheat, and barley.

On the debit side, corn requires tremendous amounts of fertilizer because it is such a heavy feeder; it needs a great deal of cultivation (or herbicides) and, especially without crop rotation, pesticides; and row crops such as corn leave the soil more vulnerable to erosion. The concept of cheap corn needs to be reexamined, not only in view of the cost increases of fossil fuels and agricultural products derived from them such as fertilizers, herbicides, and pesticides, but also in terms of its effect on the soil and ecology. Furthermore, agribusiness has paid little heed to the economics of the actual feed value of crops such as corn. It is said, for example, that protein content of corn has been dropping steadily with increased yields, with hybridization, and with chemical fertilization. This slack must be taken up by protein supplements, as must the other shortcomings of corn be met by other sources.

And yet, there is something to be said for corn in the homestead hog operation. Grain is of decided importance in the hog ration, and corn is the easiest of grains to grow on a small scale. A small plot can be planted, tended, and harvested by hand much more easily than any other grain. In fact, the new, inexpensive one-row planters now available to a gardener allow planting corn at the proper depth and spacing as fast as one can walk. A rotary tiller and a hoe, standard homestead tools, take care of cultivation. Corn can be picked by hand with much less effort than is required for the small grains; for the latter, scythes with cradles are just about impossible to come by these days, while corn harvesting requires no tools whatsoever. Small amounts of corn can be shelled with various shellers now on the market, or it can even be fed on the cob if necessary or expedient. Contrast that with threshing small grains.

Corn drying is a major expense for large producers in certain years, depending on weather conditions. And since drying is generally accomplished with LP gas, it isn't going to get any cheaper. But drying is made necessary only by technology, which includes picker-shellers or combines and massive acreages of corn which are made possible by, and are necessary to economically justify, giant machinery. Homesteaders simply leave the corn in the field until it's dry enough to store well. And, unlike some of the other crops Baker mentions, corn does store well. Pumpkins and root crops will freeze and/or rot without proper storage.

Many of the objections to corn on a commercial basis can be overcome with ease by homesteaders. They can, first of all, rotate crops much more easily than can corn farmers who run seven hundred acres; six hundred of it in corn. While big farmers can and do use organic fertilizers and practice organic methods, it is somewhat easier for homesteaders, especially those with a good supply of hog and other manures for compost. (It's always wise to compost manure rather than apply it directly to the land, since much of the value is lost in direct application under most conditions.)

With crop rotation and good fertility, weed and insect problems are minimized; this lessens the need for herbicides and pesticides. The end result is not only corn that is organic, that uses a minimum of nonrenewable resources, and that does minimum damage to the earth, but corn that is higher in food value! While we know of no definitive studies that could be called scientific, many organic farmers who have their grain analyzed for protein content (a common practice among progressive farmers of all persuasions) claim that their organically grown corn is higher in protein than that of their neighbors, or their own when they used chemicals. On a large operation this can be translated into the purchase of less soybean oil meal—and the price of this meal has more than doubled in the past few years.

Small grains

The other grains commonly fed to hogs are barley, oats, wheat, rye, and grain sorghum (milo). While these might have applications for farmers with grain drills and combines and for homesteaders who purchase grains, they will be of limited interest to homesteaders primarily because of the difficulty of hand harvesting. Small plots can be planted with a Horn Seed Sower or a Cyclone—relatively inexpensive devices operated by hand as the planter walks down the field—or they can even be planted by hand scattering the seed. The seed is then lightly raked under.

Reaping isn't that easy. The old method was to cut the ripened grain with a scythe, but that's mighty heavy work for most of us moderns. What's worse, the scythe must have a cradle to help prevent the heads from shattering and losing the grain. I don't know of any place to purchase a cradle today. Don't try to cut grain with a sickle bar mower. Like the cradleless scythe, it will cause most of the grain to fall on the ground. The only answer is a combine, which reaps and threshes the grain. But even a small one requires a good-size field to operate on, and a good-size crop to justify its expense. Small grains just aren't very practical for homestead hog raisers who want to furnish their own feed. However, for those who have the land and equipment, the small grains can be important sources of nutrition for swine.

All of the small grains should be coarsely ground for hogs. Wheat in particular is very hard if fed whole but tends to become pasty and unpalatable if ground too finely.

The importance of grinding was mentioned earlier in the discussion of Dr. Baker's *Cyclopedia,* but it could be repeated here that homesteaders with no access to grain mills should consider soaking the grains a poor substitute. It is better than feeding them whole, at least.

Barley is an excellent hog feed and is first choice in areas where corn doesn't grow. Barley has more fiber than corn and consequently more bulk, and it has slightly less energy but more protein, although the amino acid balance is not as good. Scabby (diseased) barley should not be fed to pigs, as it is unpalatable, and may even be poisonous.

Oats are high in protein, but their fiber content is too high to make a good finishing ration. They are good for breeding stock however, especially lactating sows. If oats comprise less than 30 percent of the ration, they are also acceptable for feeder pigs.

Wheat is actually superior to corn as swine feed, being equal as a source of energy and slightly higher in protein quality and quantity. Of course, it costs more than corn, largely because it yields much less per acre.

The grain sorghums are grown in the Southwest (and some other areas) where they do better than corn. They are roughly equivalent to corn in feed value and are an acceptable substitute.

Rye isn't very palatable for hogs, although it has about 90 percent of the feed value of corn. It should be limited to no more than 20 percent of the ration. Rye infested with ergot can cause abortions in sows and depressed growth rates in growing and finishing hogs, and should not be fed.

Other, less common grains can also be satisfactorily fed to swine. Rice, for example, has slightly more than 80 percent of the value of corn, but this country's total output is used as food for humans unless it is off-grade or the price is unusually low. Emmer, a rather uncommon cereal grain, closely resembles oats in composition. Triticale, like rice, is too valuable as human food to be fed to pigs except under special circumstances. Proso (or hog millet) is another pretty rare grain, but it makes good hog feed. It is nutritionally worth about 85 percent as much as corn.

Two other grains deserve mention here: buckwheat and soybeans. Neither one is really a cereal grain. Buckwheat is often grown on poor or acid soil as a late crop. It is very high in fiber; in comparison to oats, it is somewhat lower in protein and furnishes half the fat and appreciably fewer total digestible nutrients. In addition, it lacks palatability. But of even more significance, buckwheat tends to produce soft pork when it forms too large a proportion of the ration, and it also occasionally causes peculiar eruptions and painful itching on white animals, although only when they're exposed to light. If buckwheat fits in with your crop management, limit it to less than one-third of the ration.

Soybeans always intrigue homesteaders and new farmers. If soybean oil meal is so darned expensive, why can't we feed plain old soybeans? There are two good reasons. The first is that soybeans cause soft pork, just like buckwheat only worse. This is a serious defect in quality that no homesteader wants in his home-produced bacon. Soybeans should not constitute more than 10 percent of the ration, and since the protein content of raw beans is less than the protein content of processed soybean oil meal, this isn't enough to sufficiently raise the protein level of the ration.

Secondly, soybeans contain a substance called the trypsin inhibitor, or the antitrypsin factor. Trypsin is an enzyme in the pancreatic juice that helps digest protein. If the trypsin isn't there to do its job, protein is not digested, and is lost. The antitrypsin factor is destroyed by cooking and is not present in soybean oil meal.

Feeding raw soybeans to swine will not result in an appreciably higher level of protein in the diet, and might result in inferior quality, soft pork. Properly cooked soybeans are a different matter, for the antitrypsin factor is then destroyed, and the protein is nearly equal to the protein in milk or fish meal. Too much cooking, however, destroys certain valuable amino acids, which reduces the value of the protein. Cooking does not reduce the softening effect on the carcass.

Providing good pasture

While grains supply the lion's share of the hog ration (there is little doubt that most pork reaching market today was produced on nothing but corn and a little vitamin-mineral-protein supplement), other sources of nutrition also demand attention. One of the most important is pasture. We have seen the importance attached to pasture from the early 1900s (and presumably long before that) to the present. According to Morrison, nothing is more important than pasture in reducing the cost of pork production and in preventing nutritive deficiencies. "Wherever possible, swine should be provided with good pasture during the growing season and with well-cured legume hay when pasture is not available."

Alfalfa is tops for pasture and hay, followed by clover. But rape is almost as valuable as alfalfa, even though it is not a legume. The cost of establishing a stand of rape is much lower than for alfalfa.

As noted elsewhere, swine consume but little of forages, yet their importance is tremendous. Supplying their benefits on the homestead can take various forms. While pasture is the best way, because it also gives pigs access to soil and the unidentified benefits it provides, it presents problems. It is seldom economic to fence in a pasture for one or two pigs or to train them to electric fencing. The pigs will trample and waste as much as they consume, or more. Several pastures are needed, really, to allow regrowth of vegetation. Parasites can be a problem on heavily used pastures.

The answer for most homesteads, and even small farms, is to bring the forage to the pigs, along with buckets or clumps of sod from some uncontaminated source. Fields fertilized with pig manure would be considered contaminated, as well as fields that have received applications of chemical fertilizers. Get the soil from the woodlot or a remote and neglected fence row.

When I homesteaded one acre and raised one pig, we had alfalfa planted on the front lawn. When it was short I cut it with the lawn mower with a grass catcher, and fed it to the pig. The lawn looked funny of course, because I just cut enough every day for the pig, but I always did prefer a nice pig to a nice lawn. When it got higher I cut it better with a scythe, although a butcher knife would undoubtedly work as well. On our present operation we use a tractor, a forage chopper, and a wagon and get the same results. Small patches of alfalfa or clover are easy to sow by hand or with a grass seeder. However, don't plant alfalfa this spring and expect it to feed your pig this summer! Alfalfa, and clover too, is commonly seeded with a "nurse crop" such as oats or other small grains, because it grows so slowly and can't compete well with weeds. The grain is harvested and the legume doesn't produce a crop until the second year. Then, depending upon climate and weather conditions, the variety of seed planted, and other factors, it should produce two, three, or even four cuttings per year for the next three to seven years. A decent yield for alfalfa in the midland growing regions is five tons per acre. At that rate, it doesn't take a heap of land to keep a hog happy.

But here's a new twist to forage, one you won't find in any government publication. There's a crop ideally suited to homesteads, and it has a great potential for small farms too. Grow comfrey.

Controversial comfrey

Don't bother checking this out with your county agent. He might find the December 1972 release in his files attributed to Richard H. Hart, agronomist with the Light and Plant Growth Laboratory at the U.S. Agricultural Research Center in Beltsville, Maryland—which personally I find woefully inadequate and mistaken. That, most likely, will be the extent of his knowledge.

My enthusiasm for comfrey pretty closely parallels Dr. Baker's zest for pasture. I don't know for sure what's going on, but by golly I can see the results.

Let's back up a bit, in case you're not already familiar with this plant. Comfrey is a perennial of the order *Boraginacea,* the result of a cross between *Symphytum officinale* and *S. assperrimum.* It is a semisterile hybrid because of a malformed flower structure, and according to the Henry Doubleday Research Association of England—the only outfit that knows beans about comfrey—"seed is so rarely set that it cannot be supplied by any seedsmen in the world." Statements from agribusiness people on this side of the ocean commonly interpret this as, "I am not convinced that this crop has an agricultural potential because the crop is difficult to establish from seed."

Comfrey grows up to five feet high, branching out with hairy stems and broad leaves. The root system is fleshy and extensive, going down eight to ten feet and making the plant relatively drought resistant. The organic gardener will also recognize the value of the plant nutrients brought up from that depth. This crop is generally adapted to cool, moist places.

Comfrey is controversial for many reasons, paramount among them being claims for yields and nutritive values which are of obvious importance. The Henry Doubleday Research Association claims that the world record is held by J. McInnes of Kenya, with a yield of 124 tons per acre harvested in 1955. USDA spokesman Richard Hart claims that the highest yield in Vermont was a mere 37 tons per acre, and in Wisconsin 33.7 tons per acre. But considering that the yield of alfalfa in Wisconsin is 5 tons per acre, I for one am not going to haggle about the yield of comfrey. Thirty-three tons may be a long way from 124 tons, but it's also way ahead of 5 tons.

One of the more interesting aspects of this entire episode is that the yield of 33.7 tons per acre in Wisconsin was supposedly measured at a government experimental station. But an extensive check with Wisconsin agricultural officials revealed that the University Experimental Farm at Arlington planted "some" comfrey a number of years ago, but "as a curiosity . . . something for people to look at." I talked with local USDA officials, the Wisconsin Department of Agriculture, and the University of Wisconsin's agronomy, horticulture, botany, biochemistry, plant industry, and animal sciences departments—none of whom knew anything about the test plot. The source of Hart's information remains a mystery. So much for that.

Hart also downed comfrey on feed value. "Crude protein content of comfrey in the experiments discussed here ranged from 12 to 26 percent. However, digestibility of comfrey protein was only 38 and 49 percent in two tests compared with alfalfa protein which is over 70 percent digestible."

On the other hand, Lawrence Hills, secretary-director of the Henry Doubleday Research Association, said, "The average analysis of fresh comfrey is 3.4 percent protein which is the same as red clover, but with only 25 percent of the fiber so the crop is far better suited to pigs and poultry than to grazing animals which digest fiber with the help of bacteria in their extra stomachs." And researchers in Japan said the total digestible nutrients (TDN) of comfrey was not known (such tests are intricate and expensive), but they got better results from feeding comfrey than they got with ladino clover.

Which brings us back to my unbounded enthusiasm for comfrey as pig feed. In the first place, I consider it an ideal homestead plant, for reasons the USDA would never think to consider. They are interested in profitability and adaption to automation; I rather favor looking at these other factors. Comfrey is easily grown on a small scale, much more easily than alfalfa or clover. The best way to harvest it is with a butcher knife or machete, a system I still use for a hundred hogs and more. You can get a crop the first year. It's not only a perennial, but it can actually become a weed, albeit an easily eradicated one. It's a very attractive plant and could well be grown in borders and flower beds. (Can you imagine the USDA exclaiming over *that* quality in a forage plant? It could be important to homesteaders, though.)

While all of this might be of interest and importance to homesteaders, and even small farmers, the best is yet to come. I'll admit that my evidence is circumstantial and not scientific, just like Baker's endorsement of pasture before the discovery of the role of vitamins. But when somebody does finally give this the imprimatur of science, I'd like to get the credit for being the first to publish it.

Scientists already know that with the addition of vitamin B_{12}, the protein levels of swine rations can be reduced appreciably. In addition, most of the antibiotic supplements for swine contain not only antibiotics, but also vitamin B_{12}.

Morrison, the acknowledged nutrition authority, maintains that "the effect of this kind of supplement may obviously be due to the antibiotic and also to vitamin B_{12}."

Now get this: comfrey is the only land plant that contains vitamin B_{12}.

This vitamin is one of the most recently discovered and is commonly supplied in commercial rations by animal protein such as tankage, meat scraps, fish meal, and fish solubles. It is of benefit to humans and other animals afflicted with pernicious anemia. Its relationship to protein needs is interesting to homesteaders, as is its entire background as one of the "unidentified factors" in nutrition until quite recently.

It should also be pointed out here that young pigs on pasture apparently have no deficiency of this vitamin, even though plants are believed to contain little or no B_{12}—except comfrey. The fact that it has been isolated in comfrey should enhance the value of the plant.

Comfrey is not propagated from seed, but by root cuttings or roots and crowns. Cuttings will produce moderately the first year but won't come into full production for three years. They are planted between three and six inches deep, in a horizontal position, with plants spaced three feet apart. Comfrey likes a pH of 6.0 to 7.0. Weeds must be kept down the first year by cultivation, and the crop will do better with an application of nitrogen.

In northern areas comfrey begins to grow in the spring, long before alfalfa is even thinking about it. Although mature plants can reach five feet in height, they become too coarse, especially for pigs. Also, the nutritive value goes down when the plant has blossomed. For pigs, and especially young pigs, cut comfrey when it's about a foot high. Depending on soil and weather conditions, each plant can be cut every ten to twenty days. Comfrey is hardy to -40° F., according to the Henry Doubleday Research Association, and it is not troubled by insects or disease.

Once the plants have matured, the one-hog homestead

should be able to grow enough comfrey on less than 100 square feet. That would produce about two dozen plants. (Sneak out a few small leaves on occasion for yourself: to nibble on while you feed the pig, to add variety to a salad, to mix with pineapple juice in the blender for "greendrink" or to make comfrey tea.)

I don't claim to be a nutritionist. I don't pretend to know why comfrey is good hog feed. All I know is my pigs of all ages love it, and the young ones especially slicken up like fat little pork sausages when they get their daily ration of comfrey. The homesteader can add to that the ease of growing it (compared with alfalfa and clover); the low cost in terms of time, equipment, cash, and longevity of stand; and particularly the ease of harvesting and feeding. Especially if you decide not to purchase antibiotic-vitamin B_{12} supplement, comfrey just makes a lot of sense.

If it's easier for the farmer with equipment to grow grains, the worm turns when it comes to roots and other succulents in addition to comfrey. The homesteader has no disadvantage at all here.

Root crops

Baker advises that "Artichokes, potatoes, ruta-bagas, parsnips, carrots, and beets, are readily eaten by swine, and are preferred in the order named. We have successfully wintered hogs entirely on ruta-bagas with the addition of a little meal. Artichokes are a cheap and excellent root food, if the swine are allowed to gather them themselves in the autumn and spring." The artichokes he refers to, of course, are Jerusalem artichokes, which are related to the sunflower and produce potatolike tubers.

Although, as Baker indicates, roots were once quite common and are still used to a great extent in northern Europe, this is a generally insignificant source of livestock feed in the United States.

One reason is that root production requires a great deal of hand labor (or extremely specialized and expensive machines not suitable for producing livestock feed), and roots are much harder to store and feed than is grain. Roots just don't lend themselves to the big-business kind of farming practiced in the United States. Climatic differences between northern Europe and the United States also play a role, but even where corn and the sorghums don't thrive in this country, roots are of little economic importance.

With these kinds of facts—and bearing in mind the primary goals of American agriculture—agricultural experts can't recommend roots for livestock. They don't even think about them, and if you ask you'll either confuse or amuse them. But homesteaders march to the beat of a different drummer.

We can't say homesteaders want crops that require a great deal of hand labor; homesteaders are just as smart, and perhaps almost as lazy, as anyone else. But we can quite definitely maintain that homesteaders do *not* want crops that require expensive machinery. Grain can be grown with hand tools, but it takes a lot of work. Roots can be grown with hand tools, and they don't take quite as much work. If you have swine and smarts, it might not take very much work at all.

Artichokes are a fine example. According to some sources they yield up to twenty tons per acre. I can't find anyone to verify this, but I do know that in southern Wisconsin they yield a lot more than potatoes do. Artichokes are planted like potatoes, and no doubt for larger plots the various types of potato planters could be used. But you can also plant quite an area with a furrowing attachment on a rotary tiller or garden tractor, or even a hoe if need be. Drop the eyes in and cover them up.

They can be mulched if you have the materials. I don't, and I don't have the time. So I run down each row once with the tiller when the weeds come up. After that the artichokes are so thick and tall that any weed growth is stifled.

Then comes the best part. I used to go out with a fork and

dig the tubers at harvesttime, which soon made cleaning the Augean stables seem like changing the newspaper in the bird cage. The solution is in the quote from Baker just mentioned: let the hogs do the harvesting. They're better at digging than you are anyway, they have more time for it, and they certainly enjoy it more.

It's been said that if hogs are left on a patch of artichokes just long enough to clean up most but not all of them, and are then removed, you won't even have to replant the next year. While I can't personally vouch for the system, it sounds plausible; indeed if it works, it would be the epitome of efficiency.

The other root crops mentioned can also be grown specially for hogs, although the average serious homestead should have sufficient culls and surplus to feed one or two pigs quite well.

Vine crops

Pumpkins are interesting. Baker said there is no better fall food than pumpkins and grain boiled together. And pigs love them even without grain and without cooking. Remember a few chapters back where we spoke of the importance of swine in the orient? Well, pumpkins are just about as important. Pumpkins are used in China about the way we use potatoes.

While pumpkin is very rich in vitamin A, this isn't of great importance to hogs that have access to pasture or well-cured hay. And, while none of the other nutritive values of pumpkin are particularly thrilling either, there is one minor point that will interest homesteaders and organic farmers—pumpkin seeds are said to eliminate prostate trouble, for some unknown reason. This might be of no practical value at all, and yet it's intriguing to put it all together. Pumpkins were considered good hog feed in the United States in 1900; scientists acknowledge that swine require unidentified vitamins and factors, and pumpkins apparently have unidentified factors; the Chinese—whose acupuncture and other ideas we laughed at a few years ago but which we don't laugh at quite as much any more—have raised pumpkins and hogs for years.

What's it all mean? I dunno, but like I said before, modern homesteaders with heads on their shoulders have to think about *something* as they shovel manure and feed pumpkins to the hogs.

What gardening family does not have an overabundance of zucchini squash? Feed it to the pigs. That goes for other squashes too, as well as tomatoes, melons, and any other culls or surplus from the garden or orchard. If pigs, like men, are "universal feeders," anything that keeps you healthy will keep your pig healthy.

And if pigs relish grubs and whole frogs and worms, they're obviously a little less squeamish than most of us. That's why we can give them "the slops of the house" as Baker puts it.

Edible waste

Most of our garbage isn't really garbage at all: it's likely that some countries could *live* on what Americans throw out. And not just average Americans, who eat canned spaghetti and frozen pizza; on a nutritive basis, homesteads could be even worse.

The reason for this statement is that recent research has considered food discarded from only the kitchen. What about the waste before all that packaged and processed stuff even gets to the kitchen? If the homestead processes homegrown foods, the garbage will include not only Junior's uneaten spinach, but carrot leaves, potato peels, beet greens (they're delicious, but you can't eat 'em *all*), pea pods, rinds and stems and skins and seeds of all kinds that, if we were really hungry, we could live on. That may be waste, but it's not garbage. Grinding it up and sending it down the sewers to already overloaded sewage plants is more than waste—it's senseless, and sinful. Besides, the pigs will enjoy it.

You ought to know, though, that feeding garbage is illegal without cooking it at 212° F. for thirty minutes. The homesteader, and especially anyone concerned about trichinosis, should know why. Trichinosis is a parasitic disease of humans (and possibly all mammals) caused by *Trichinella spiralis,* the major source of which is infected and undercooked pork.

All parasites exist in cycles and the trichina is no exception. The adult parasite lives in the small intestines of man, hogs, rats, and other mammals, and the female produces larvae in the lining of the intestines. The larvae go through the wall of the intestines into the lymph stream, the blood stream, and into the muscle cells where they grow and curl up into a spiral that becomes encapsulated. They can stay there for years. To complete the cycle, the muscle tissue must be eaten by another animal (man or beast). Cooking the meat to at least 170° F. kills the parasite, however.

Therefore, in order for a human to get trichinosis he must eat meat that is infected with it, and that meat must be raw or inadequately cooked. That's the only way a pig can get it too, of course. So humans cook their pork and pass laws that, for all practical purposes, dictate that lest some infected and uncooked pork gets through, all garbage must be buried in sanitary landfills, incinerated, or otherwise wasted.

The homesteader should adequately cook his pork, just like everybody else. He should see to it that no raw or undercooked pork, or better that no pork at all, gets fed to the pigs. And then he should feed the potato peelings and beet greens and other items, that could just as well make soup stock, to the pigs. There is no possibility of such "garbage" producing trichinosis.

(It should be noted that trichinosis is possible without feeding the pigs raw pork, however. See the section on diseases for more details.)

Dairy products

The last category we come to is dairy products, and depending on the homestead, this can be a major food item.

If a homestead has a cow or goats with the intention of producing enough milk and dairy products for the household on a year-round basis, the homestead is sure to have a surplus of milk at certain times of the year. The reason for this is simply that a dairy animal freshens, the milk supply increases for awhile, then tapers off and finally stops. If the home dairy is managed to produce enough milk for the family during the lean period, it follows that it will have too much during the flush period. Some people will want enough milk to make cheese, butter, yogurt, and ice cream ten or twelve months of the year, and they'll still have a surplus after the animals freshen. Other families will be content to produce only drinking milk for most of the year, and to make dairy products only when they have a lot of milk. They will still have skim, whey, and perhaps a surplus of whole milk.

In any of these cases the hog will make excellent use of what would otherwise be waste. Milk is known as nature's most perfect food because of its high quality proteins, vitamins, balance of minerals, and the beneficial effect of lactose, the milk sugar. And do pigs love it! They'll learn to recognize you coming with the bucket and they'll get so excited they'll make those "come-and-get-it" dogs in the TV dog food commercials look about as eager as mice approaching a baited trap.

Milk isn't fed on a commercial basis, but only because of its cost. In addition to the high cost of whole milk today, shipping, storing, and sanitary feeding on a commercial scale cause problems. In the old days when small creameries dotted the countryside in dairy regions, individual farmers skimmed their milk and sold only the cream, the skim going to the hogs. But now centralized plants buy whole milk, work it over, then send back to the farms such products as cheese meal, dried whey, milk sugar feed, condensed whey, dried skim milk, and others. (Today, more of these are fed to poultry than to swine.)

On the homestead, milk and milk byproducts are the most valuable feeds available. Nutritionists tell us a pig can thrive on corn and about a gallon of skimmed milk a day, so if we add comfrey and some of the other items we've covered here, how can we lose?

Once again the best is yet to come, for we run into another unidentified factor! Milk and milk byproducts hold in check some of the internal parasites of swine. This has been observed, and has also been backed up by research. But even scientists don't know why or how. That doesn't really matter to homesteaders who feed milk to eliminate the need for tankage or fish meal and get an "organic vermifuge" in the bargain.

Skim milk is higher in protein than whole milk, and has about twice the protein of whey. All are low in vitamin D and iron. Skim milk is the best possible protein source for swine, especially young swine.

A hog should get about a gallon to a gallon and one-half of milk per day. While this amount will be a smaller percentage of the ration as the pig grows and eats more, protein needs also decrease then.

From the information presented in this chapter we can see how and why the economics of big farming have changed the diet of the nation's pigs, and why the agribusiness experts tell us we can't raise pigs the old-fashioned way. They just haven't thought enough about it, and they aren't attuned to the special needs, the special goals, and the special capabilities of the small farmers or homesteaders. Homesteaders want to know why they have to feed hogs concentrates and supplements and antibiotics and vermifuges, and the only answer agribusiness can give is a condescending smile. There *is* an answer. They don't have to.

Not only does following old-fashioned methods allow homesteaders to compete with big-time operators. It also allows them to produce pork that tastes as good as pork did in 1900!

Outline of one year's feeding management

We have seen that homestead hog feeding can be, and probably *should* be, considerably different from the commonly accepted commercial methods in vogue today. This calls for another look at overall homestead hog management.

The question often arises as to the best time to raise a pig. From the information already discussed here, there would seem to be no doubt about it: in temperate climates at least, buy a weanling pig in spring. There are so many reasons for this, and they make a pig blend in so smoothly with the entire cycle of homestead activity that a homestead without a hog is like a puppy without a tail to wag. It can function, but something's missing.

It can begin with the milk flow that's so important to getting baby pigs off to a good start. If the homestead raises goats, most of them freshen in late winter and early spring because of the seasonal nature of estrus (heat) in those creatures. Milk flow will then peak in April, May, or June. Even if the home dairy animal is a cow that has freshened at some other period, milk flow is likely to increase when she is turned onto lush green pasture in late spring. If you can come up with a gallon of milk a day for your new pig, you have it made.

At that time of the year the comfrey, with its important vitamin B_{12}, protein, and other valuable components, will be young and tender and just right for baby pigs. Of course, so will the alfalfa, clover, pigweed, lamb's quarter, and other forages.

Grain stored the previous fall will still be in good condition, and in the spring it's much easier to estimate how much of what grains are available. On our place anyway, we know how much grain we have in the fall and make up rations for the cow, goats, chickens, ducks, rabbits, and other livestock accordingly. But by spring there is always too much of one left over and not enough of another. Since grains have different qualities, we must combine different elements to feed a balanced ration depending on the grains used.

Temperatures that are too hot or too cold result in an increase in the amount of feed needed to produce a pound of meat. It's easier to keep a pig cool in summer than it is to keep it warm in winter, and besides, more and cheaper feed is available in the summer. Housing can be simpler and less expensive if it's used just during the milder part of the year.

The pig will grow rapidly during the summer—just as the garden will. This means not only that the pig will get thinned carrots, beets, and similar vegetables to eat, but overgrown lettuce, pea pods and vines, and similar early crop residue. In addition, as the harvest season begins, the wastes of the kitchen will mount up: as the homesteader preserves beans, broccoli, and other vegetables on through the season, more and more trimmings and culls will be available as more crops ripen, and the growing pig will eat more too.

By the time the pig reaches the fattening stage, the garden will be ready with pumpkins, artichokes, potatoes, rutabagas, squash, and turnips. Whether grown specifically for the pig or whether only culls and surplus go into the trough, there is no time of the year when symbiosis is more gratifying. There is a beautiful web that could almost symbolize homesteading: the garden feeds the homesteader and the pig, and the pig produces meat for the homesteader and manure for the garden.

The grains also ripen at this time of the year, and even if the product is only a little corn, it can be fed right from the field without the work of storage and extra handling. Then the pig can be turned in to do the gleaning, of the grain fields and the garden, or perhaps in some areas the woods that will yield acorns and other mast.

And finally, on a frosty October morning, the cycle is completed. The pig is butchered. At this time of the year there is no pressure from newborn kids and lambs, no rush to plant or cultivate or mulch or harvest or preserve, no demands of tiny chicks nor rush to make hay between rain clouds. There are no flies. The nighttime temperature that dips toward freezing is ideal for the home butcher, who doesn't have the luxury of a walk-in cooler.

The lard is rendered—not on a steaming summer day nor one in blustery winter when the discomfort of open windows is no better than the fumes of rendering lard—but on a day that makes use of nature's air conditioning. It's traditional to make doughnuts with freshly rendered lard, and what better time is there to eat doughnuts than on a crisp fall morning?

Then, just about the time the bacons are cured and come out of the smokehouse, the first snow falls. The homesteader enjoys a breakfast of fresh milk, fresh eggs, and delectable home-cured bacon, and leans back in his chair. Along with whatever other provisions he has stored, with the stock in his barns and the feed in his mows and bins, he has 135 pounds (more or less) of the finest pork available. There is no concern with stretching feed to maintain another animal, no worry about winter housing, no hassle with frozen water buckets on blizzardy mornings. The homestead is ready for winter.

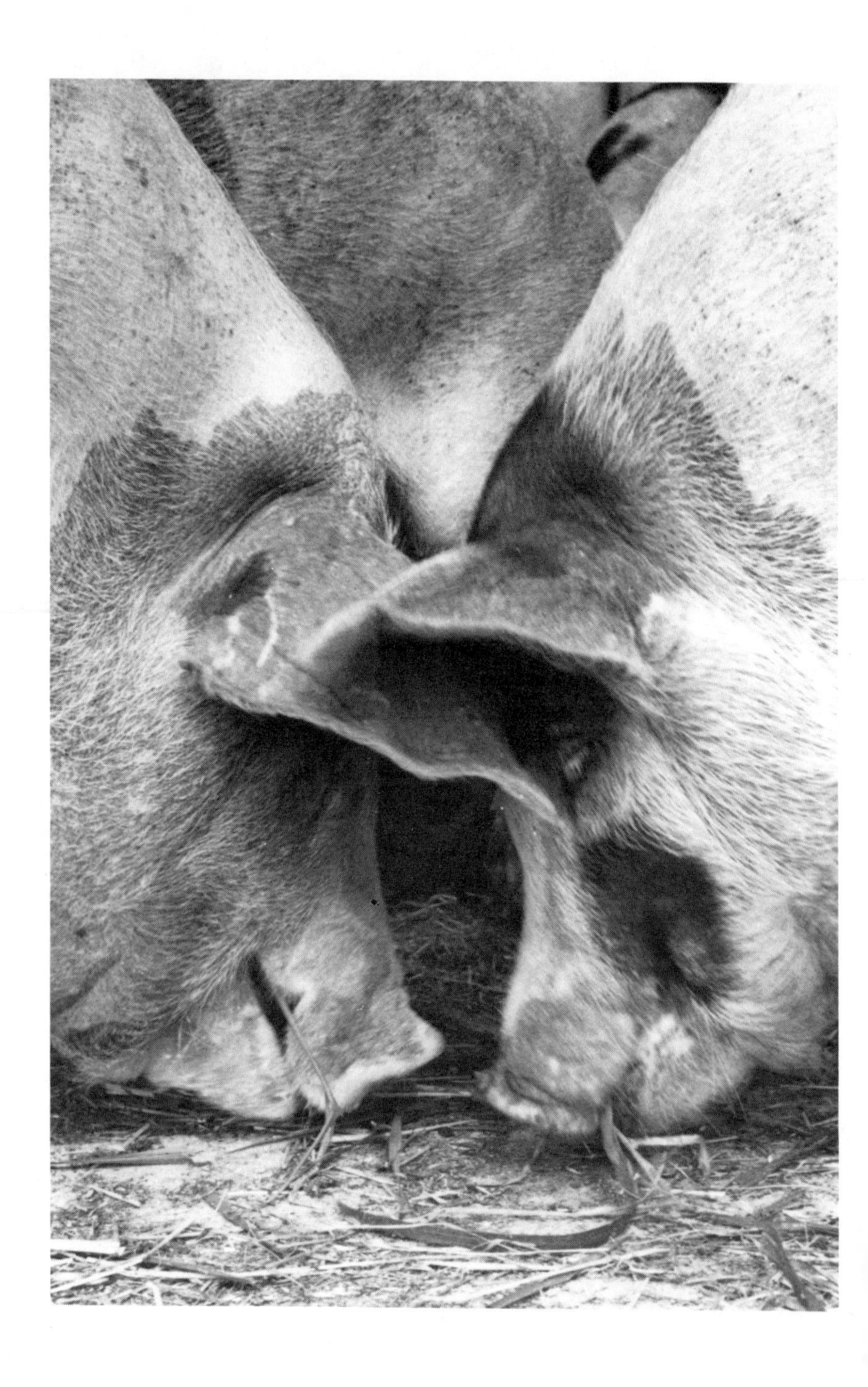

Chapter 8.
NUTRITION

The complexities of hog nutrition

We have discussed swine feeding from two different viewpoints. One is the homestead view which eschews commercial feeds and chemicals and endeavors to make use of homestead feeds. The other is the outlook of the commercial factory farmer whose primary goal is —must be— profit. Oddly enough, the homesteader must actually know more about nutrition than the commercial hog producer. The explanation is simple.

Under modern conditions and according to present theory, formulating hog feed is extremely complicated. It requires a knowledge of the nutrient levels of the various feeds available, as well as the nutrient needs of different classes of stock under different conditions. It involves use of such additives as antibiotics, arsenicals, nitrofurans, and sulfonamides. Mixing such feeds requires a great deal of technical knowledge and the use of weighing and mixing equipment few hog producers can afford. Since very small amounts of some ingredients must be accurately weighed and mixed with very large amounts of other ingredients, the process requires delicate balances that are accurate to 0.1 gram (as well as other weighing devices) and a mixer that will evenly distribute minute quantities of vitamins, etc., throughout large batches of feeds.

Even very large hog producers are not able to economically justify the purchase of separate ingredients for feed supplements, and some of them (certain vitamins) do not stand up well in storage. Only large feed manufacturers can economically supply these ingredients.

Crystalline vitamins and drugs (called micro-ingredients) must be carefully diluted before being mixed with other ingredients, and this, too, requires special equipment and skill. Other ingredients also require special equipment. One example would be waste animal fat, which entails a heated storage tank, metering pump, and blender.

Some hog feeds (especially pig starters) are in pelleted form for reasons of economics that are important to commercial hog farmers. But pelleting requires not only all the mixing and weighing equipment: it also requires a pelleting machine, which makes pellets from ground feed by steaming it and extruding it through dies.

In all, there may be as many as twenty different ingredients in a modern complete hog ration—without the grain!

If hog feeding is this complicated, why doesn't the feeder have to have advanced degrees in nutrition, physiology, chemistry, math, and a few other sciences? Because no one could have all that knowledge, spend all that money on equipment, spend all that time formulating and mixing feeds—and still have time to raise pigs! The farmer pays other experts to perform these tasks for him, and consequently if he takes their word for it, he doesn't have to know very much about it at all.

In contrast, the homesteader doesn't make use of these experts if he follows homestead feeding techniques. What's more, his methods fly in the face of advice from even the generalized hog experts! Any homesteader who seriously intends to compete with modern commercial hog-raising methods probably has a lot of studying to do, and even the casual homestead hog raiser had best learn at least some basic facts about hog nutrition.

Practical homestead hog feeding, then, combines the best of modern scientific knowledge and some of the basic practices in the old-fashioned "scoop of this and scoop of that" method. One relatively simple way of looking at this (without going into any great detail on hog nutrition) is to examine what modern technology tells us hogs need and how agribusiness fills those needs, and then find out how we can meet those needs on a homestead basis.

In the abstract, nutrition is a very simple science. All you need do is eat foods that contain proper amounts of the elements needed to sustain life and growth. But—what are those elements, what foods contain them and in what quantities, and how do they interact? In practice, nutrition obviously is *not* a simple science.

The homestead hog raiser is in a particularly difficult position because he often makes use of feeds just because they are available or cheap; because these ingredients change over the growing season; and because there is no modern research on what we're calling the homestead method of hog feeding. Anyone involved in any form of animal husbandry really should have at least a basic knowledge of nutrition, but for those who stray very far from bagged feeds with neat labels advertising protein, fat, fiber, and other important components, the need is even greater. You are what you eat, and the same holds true for a pig.

Food has two basic purposes. The first is obviously to

keep the animal or person alive. It takes energy—fuel—to pump blood, to breathe, to digest, to keep warm, and even to blink eyelids and wag tails. Food is fuel, and there are high-grade and low-grade fuels.

The second purpose of food is to provide health and growth. This is especially important for young, rapidly growing animals, for pregnant females, and for females nursing young.

The life-sustaining aspect is most important, because if animals don't have enough nutriments to stay alive, there obviously isn't any opportunity to bother with growth or reproduction. But the only reason for feeding hogs is to have them reproduce and grow, so we're concerned with a level of feeding that will do much more than just keep them alive.

Pigs need over thirty-five individual identifiable nutrients. The amounts and proportions of these vary with the ages and the demands (growth, reproduction, etc.) placed on the pigs. There is no single source for all these nutrients.

The two most important elements of feed, because they are essential to vital life processes, are energy and protein.

Calories

Energy nutrients are primarily carbohydrates and fats, although protein can also supply energy. Energy nutrients in excess of those required for vital bodily functions are stored as body fat. Energy therefore takes on added importance in the latter stages of hog raising as we "finish" the animals.

The energy value of a feed is commonly expressed in feed tables as TDN, or total digestible nutrients. Other nutritionists use DE, or digestible energy. But perhaps the best way for modern city-bred, first-time hog raisers to think of it is in terms of calories. We're all familiar with calories in our own diets, and a calorie is a unit of heat, which is energy.

Plants contain different forms of carbohydrates. On feed tags these are listed as nitrogen-free-extract (NFE) or crude fiber. NFE includes sugars, starches, and some hemicelluloses—the more soluble carbohydrates. Crude fiber is cellulose and very complex carbohydrate. Crude fiber can be utilized by ruminants such as cows and goats because of the fermentation-vat function of their four stomachs. Pigs, like people, are monogastric and do not digest crude fiber.

On the average, cereal grains contain from 60 to 70 percent NFE and are low in crude fiber, which is why they are widely used in swine feeding. TDN values are arrived at by adding up all the organic digestible nutrients—protein, fiber, NFE, and fat—times 2.25. (Fat is considered to have 2.25 times as much energy value as protein or carbohydrates.) Multiply TDN by 2,000 to convert to kilocalories of DE. Corn containing 80 percent TDN has a DE value of 1,600 K cal. per pound.

Grains are the chief source of energy for swine. While several other grains may be used, they are measured up against corn. But corn is deficient in many important nutrients. It is deficient in phosphorus, calcium, salt, vitamin A, vitamin D, B vitamins (including B_{12}, riboflavin, pantothenic acid, niacin, and choline), and the unidentified factors. It also doesn't have enough protein and the protein is not of acceptable quality.

Protein is very important not only because it is the body-building ingredient in feed but also because it plays an important role in the body regulators such as hormones and enzymes.

The recommended protein content of feed for hogs ranges from 20 percent for creep feeding and 22 percent for early weaned pigs, decreasing to 18 percent at 40 pounds; 14 percent at 80 pounds; and 13 percent at 120 pounds. Gestating sows and gilts need 14 percent and lactating sows need 15 percent protein. Corn might have 9 percent, but it is deficient in certain amino acids.

A feed can contain sufficient protein and still be deficient. Protein is a complex nutrient composed of amino acids, and the amino acids are what's important. There are at least twenty-four amino acids, but since they combine like letters of the alphabet, there could be as many proteins as there are words in the dictionary. Essential amino acids are those required by an animal but not synthesized. They must therefore be included in the ration. But looking at protein, per se, does not guarantee the presence of any given amino acid. Different animals need different amino acids, and they might consume a feed that is high in protein but deficient in the specific amino acids they require. In that case the "quality of protein" is poor.

Amino acids are needed for the formation of every new cell, so quality of protein is a basic requirement for growing pigs, and especially for young pigs. It affects not only meat (muscle) but internal organs, blood, and bone. Without the proper amino acids the animals cannot develop properly.

When the swine ration consists of grain, which is deficient in crude protein as well as essential amino acids, the main problem of balancing a ration centers around correcting the amino acid deficiencies with protein supplements. It is generally acknowledged that the hog diet should consist of protein derived from both plant and animal sources, so common protein supplements include soybean oil meal and tankage, for example. On the homestead, milk and milk byproducts may be the most important protein sources.

The most common supplement used to increase the protein level of corn-based rations for swine is soybean oil meal, which is 44 percent protein. (Because the oil meal is processed, it does not cause soft pork like raw or cooked soybeans do.)

The homesteader can buy soybean oil meal and still have organic pork, of course. But if the goal is complete self-sufficiency, homestead protein sources include milk (one gallon a day), legume forage or rape (which is not a legume but which equals alfalfa in protein content for hog feeding), and other grains which are higher in protein than corn is.

The following amino acid requirements have been determined for weanling pigs, expressed in percent of total ration:

L-Arginine	0.20
L-Histidine	0.20
L-Isoleucine	0.70
L-Leucine	0.60
L-Lysine	1.00
DL-Methionine	0.22
DL-Phenylalanine......	0.46
L-Threonine	0.40
DL-Tryptophan........	0.20
L. Valine	0.40

Soybean meal and fish meal are good sources of these essential amino acids. But homesteaders can be assured that their feeding practices are scientifically sound too, by remembering Morrison's statement: "There is not apt to be a deficiency of any amino acid in a ration where the protein comes from three or more good sources, and where a considerable part is supplied by such feeds as soybean oil meal, fish meal, or dairy by-products. This is especially true when alfalfa hay or alfalfa meal is included in rations for pigs not on pasture."

In other words, with a feeding program that includes milk, vegetables, and pasture, in addition to grain, amino acid deficiencies are unlikely to occur.

Low levels of protein tend to produce more fat and less lean in the carcass, which is undesirable. Low protein levels result in slow and expensive gains. But conversely, protein is expensive and high levels of protein are simply wasted. Without souped-up protein concentrates the homesteader is more likely to err on the low side, and therefore should be aware of protein requirements and levels.

The safest course involves giving pigs access to *good* pasture or feeding legume forages; allowing a gallon of milk or milk byproducts per day per hog; feeding a variety of grains instead of just corn; and making use of other feed sources such as vegetables and root crops. Barley, wheat, oats, and rye all have more protein than corn.

Minerals

So far, feed formulation is still relatively simple. It's a matter of finding the nutrient requirements of pigs and the nutrient composition of the feeds available, and doing a little arithmetic. However, it also includes such factors as nutrient interactions, cost, palatability, use of additives, variations in nutrient content of feeds in local areas during different growing seasons, and more. And then there is the matter of vitamins and minerals, which complicate the arithmetic even more.

Of the minerals that make up the body (and which therefore are of obvious importance for health and growth), calcium and phosphorus account for more than 70 percent. A deficiency of one or the other can result in poor gains, rickets, broken bones, or posterior paralysis. A large excess of either can interfere with the absorption of the other, so too much can have the same effect as too little. Swine need 1.2 to 1.5 times as much calcium as phosphorus. Since grains have more phosphorus than calcium and grains are a staple in the hog diet, swine are more commonly deficient in calcium than in phosphorus. However, this is further complicated by the fact that more than half of the phosphorus in grains is in the form of phytin phosphorus, a form which is poorly utilized by hogs.

Iron and copper are necessary for hemoglobin formation and for the prevention of nutritional anemia. Since sow's milk is very low in iron and copper, suckling pigs are especially prone to be deficient in these minerals. Once pigs begin to consume feeds other than milk they usually get enough iron and copper, especially if they have access to pasture.

Zinc is an important mineral for hogs because a zinc deficiency can cause parakeratosis, poor growth, and low feed efficiency. High levels of calcium and/or phosphorus in the diet appear to result in a higher zinc requirement.

Sodium and chlorine occur in the fluids and soft tissues of the body and play vital roles. Salt contains both, and should comprise 0.5 percent of the ration. Salt deficiency results in slower gains. Swine need less salt than do other classes of livestock. Usually one-half pound of salt or less is mixed with a hundred pounds of feed, but it can also be fed free-choice. By using a trace-mineralized salt, the feeder is also insuring against possible deficiencies in iron, copper, manganese, and iodine. These are not always supplied even by organically grown feeds raised on fertile soils.

Iodine is essential, and may be deficient in the feeds grown on certain soils. Iodized salt takes care of the matter.

Commercially prepared rations are likely to contain ground limestone for calcium, and dicalcium phosphate, defluorinated phosphate, steamed bone meal, or other low fluorine phosphate material to supply readily available phosphorus. But here again, homestead feeds such as milk and legume forage or rape are rich in calcium, and tests have shown that the phosphorus from grains and other plant sources is adequate for satisfactory growth.

Other minerals needed by swine are found in sufficient amounts in natural foodstuffs. These include cobalt, magnesium, manganese, potassium, sulfur, and selenium.

Vitamins

The other major component of feeds is vitamins. Vitamin A occurs as carotene in plants and is converted by the animals. Yellow corn is a source, but an unreliable one. Lush pastures or green leafy hay are usual sources of vitamin A, and pigs on pasture are adequately supplied. However, vitamin A is usually added to swine rations, and it's necessary for hogs fed grain and supplements.

Vitamin D is essential; but except for hogs raised in confinement, enough comes from sunshine. Vitamin D is needed for assimilation of calcium and phosphorus, and therefore is essential for normal calcification of growing bone. It is supplied in commercial rations by adding irradiated yeast, but if the hogs are allowed outside, the ultraviolet rays of the sun provide plenty of vitamin D even in the winter.

The B vitamins are added by means of animal or marine proteins such as tankage and condensed fish solubles, and in distiller's dried solubles. Increasingly, the pure vitamins are added. But once again, the pasture scoffed at by modern, progressive hog farmers takes care of the situation in nature's way.

There is one exception. Vitamin B_{12} is found in no feeds of plant origin except comfrey. (It should also be noted that, while alfalfa does not contain B_{12}, hogs on alfalfa or other good pasture are not deficient in this vitamin.) Vitamin B_{12} is especially important for young pigs; and, by inference, comfrey is especially important for young pigs.

The following B vitamins are also important:

• **Riboflavin (B_2).** This water-soluble, B-complex vitamin functions in the body as a constituent of several enzyme systems. Therefore a deficiency results in a wide variety of symptoms including loss of appetite, stiffness, dermatitis, and eye problems. Pigs may be born dead or too weak to survive, or poor conception may result. Requirements of riboflavin appear to be higher at lower temperatures. A range of 1.0 to 1.5 mg. per pound of ration is recommended, but cereal grains are poor sources.

• **Niacin.** This plays an important part in metabolism as a constituent of two coenzymes. Loss of weight, diarrhea, and dermatitis are common deficiency symptoms. The niacin in cereal grains may not be available to swine because it is in "bound" form, but on the other hand the amino acid tryptophan can be converted to niacin. The National Research Council recommends 5 to 10 mg. of niacin per pound of feed for growing pigs.

• **Pantothenic acid.** This B-complex vitamin is a constituent of coenzyme A which functions in cellular oxidation of food materials. That sounds important, and it is, because it affects growth and health. The requirement is 5 to 6 mg. per pound of feed.

• **B_{12}.** This is one of my personal favorites, because it was discovered only in 1948 and even today nobody knows for sure just what it does. Not that I delight in ignorance—I just think humility is a fine attribute, even for scientists. We do know that B_{12} is essential for metabolism and that grains and plant products, except comfrey, are poor sources. Only 0.01 mg. per pound of feed is required, so it's not the sort of thing you shovel into the feed trough.

Other important vitamins are either present in normal rations or synthesized by the pigs: vitamins C, E, K, thiamine, pyridoxine (B_6), choline, and biotin. The last word isn't in on these either, though. For example, recent research indicates that additions of choline to sow rations may be beneficial.

We should discuss antibiotics and other additives, if only because they are so widely used in swine rations. They are not nutrients of course, but they have some of the same effects.

Antibiotics

Antibiotics, for instance, generally increase average daily weight gains, improve feed efficiency (meaning it takes less feed to produce a pound of pork), improve uniformity of performance (which is important to commercial producers), and may reduce death losses during the growing period. No one knows for sure how they do all this, although they apparently have an influence on intestinal bacteria. For starter rations, forty grams per ton of feed is the usual dosage; twenty grams per ton are used from weaning to about one hundred pounds; and ten grams per ton after that. They are not beneficial to breeding animals. There are some indications that antibiotics improve the health of unthrifty pigs or that they control bacteria which compete with the animals for vitamins and other nutrients. Healthy pigs respond less to antibiotic feeding. The organic feeling, of course, is that animals fed antibiotics routinely have less natural resistance, and that humans consuming the resulting meat are also eating antibiotics and are therefore reducing their natural resistance. The question seems almost academic to the homestead hog raiser, because antibiotics are almost impossible to add to home-mixed rations, and have no value under homestead management anyway. The same can be said of antimicrobial compounds, which inhibit specific harmful microorganisms in swine and improve feed and growth efficiency. Arsenical and nitrofuran compounds are widely used in commercial premixes.

Research is being done on hormones and enzymes, but thus far these do not show much practical value for swine so even the commercial producers don't use them.

The details in this chapter probably don't have much meaning for homesteaders who don't have a laboratory in the basement and a couple of college degrees to go with it. But the basic message is *very* important. That is, all food is not the same. A pig cannot live on corn alone, much less thrive—even corn and pasture or barley and soybean oil meal won't do the job. Certain basic nutritional requirements must be met. Our forefathers did a pretty fair job through experience, intuition, and good sense, even without a great deal of technical knowledge. We know more than they did about certain technical aspects even if we aren't scientists or nutritional experts. Our problem as homesteaders is acquiring their experience, using our common sense, and combining that with what we know of the scientific aspects.

With that course, who knows? We might come out ahead of both grandpa *and* the purely scientific experts!

Then we come to the "unidentified factors," which are still controversial. As noted, these seem to be present in soil primarily, but also to some extent in forage (especially comfrey) and milk.

Hogs are well known for their ability to balance their own diets if given sufficient opportunity by your providing a choice of feeds. Neither will they make hogs of themselves and overeat, unlike most other farm animals. These two factors, along with the principle of using feeds from several different origins, make homestead hog feeding relatively foolproof even without in-depth knowledge.

Remember also that growing and finishing pigs don't make very efficient use of forage, important as it is for nutrition. Grain is a necessity, although roots and vegetables can replace a portion of it.

A model feeding program

In line with what has been said above, the homesteader who purchases a weaned pig in the spring might follow this feeding program:

The young pig should get as much comfrey as it will eat and at least a gallon of skim milk, buttermilk, or whole milk per day. In addition the pig should have access to ground grains, and preferably a selection of them such as corn, barley, wheat, and oats. You can figure that a pig will eat from four to six pounds of feed per hundred pounds live weight. Your weanling pig of 40 to 50 pounds will probably eat about six hundred pounds of concentrate to reach a slaughter weight of about 220 pounds.

Give the pig kitchen wastes (not poultry bones or raw pork), cull, and excess garden and orchard produce.

As root crops, pumpkins, and other crops mature, these can replace more of the grain.

Some producers prefer to limit feed after the hog reaches 120 pounds. This increases feed efficiency, and is said to improve carcass quality or the lean-to-fat ratio. Limited feeding can be estimated at 70 to 90 percent of full feed, or the amount eaten in twenty to thirty minutes. This should come to about one pound of feed for each thirty pounds live weight.

Feeding swine organically by homestead methods is less efficient during the winter, and more difficult, especially if milk is in short supply. Milk is a major source of nutrition in the plan just outlined, and a gallon a day is a lot of milk for many homesteads, even during the summer. Comfrey, and pasture in general, is also very important in homestead hog raising, and if these are available at all in cold climates, they will be in dry form and therefore of less value. And of course, there will not be the abundance of vegetables and other summer feeds.

If the pig is fed only grain, and particularly if the ration consists of a single grain, the homesteader had best follow commercial feeding practices by including a protein supplement. This will replace the protein and calcium of milk with soybean oil meal, tankage, and similar products. Of course it will also include vitamins (which is all right) and antibiotics (which isn't all right).

Even the best alternatives are inefficient. Pasture can be replaced, in part at least, by well-cured legume hay and by cured comfrey. Both of these obviously entail more labor and/or expense than the fresh product, with less value. Roots and pumpkins can be stored and fed to supplement grains, but they deteriorate in storage, and storage involves double handling.

While it shouldn't make a difference in actual practice, there is less sunshine (vitamin D) in winter in many locations. And lurking behind all of these procedures is the fact that the pig will require a great deal of feed just to keep warm and the fact that you'll be battling frozen water and snowdrifts to do the chores.

Then, depending on the timing, you'll be faced with the job of butchering just at the time you should be getting into all the spring homestead chores! Some people like to raise their pig during the winter, but not me.

We could go into much more detail on the nutrition of hogs: the various nutrients they need at different life stages and the nutrition available in various feeds. Important as these factors are, they're much too complicated to be of practical use for homesteaders. Moreover, homestead feeding that includes milk or milk byproducts, pasture, comfrey, a *variety* of grains—a variety of ingredients in general—is almost certain to be successful.

Nutritional deficiencies

According to some sources, nutritional deficiencies among hogs are becoming increasingly common. A little introspection will provide some clues as to why.

One swine authority, writing in 1952, cited "leached and

depleted soils'' as one of the reasons for nutritional deficiencies, and he was not a spokesman for organic farming. The problem is even more serious today, and will be considered especially so by those convinced of the efficacy of organic methods.

Therefore, one of the most important considerations in feeding livestock of any kind (and yourself too) is to use feed that was grown on fertile soil. Most homesteaders and small farmers are vitally interested in their soil, but building a healthy and productive soil on most of our abused farmlands is a long and arduous task. Pigs, of course, are of great importance in this concern because they are among the most prolific providers of barnyard manure.

However, remember when we said that chemical farmers *must* use more chemicals to get a crop, while organic farmers are able to rely less and less on chemicals every year? The same endless circle is in evidence here; while hogs need feed grown on good soil, the feed grown on poor soil also produces less valuable manure! The rich get richer and the poor get poorer.

Chemically fertilized crops are not the same as those grown on naturally fertile soil, even though agribusiness experts like to point out that a plant can't tell the difference. This isn't the place to investigate that statement, but if you aren't convinced, consider for a moment that virtually all hog authorities believe the ''unidentified vitamins and factors'' come from soil. There is too much we don't know, to enable us to stray very far from nature's ways without a great deal of trepidation. At any rate, good feeds come from good soil, and deficient soils can only produce deficient feeds.

Forced production is another cause of nutritional deficiencies in swine. Very rapid growth, one of the major goals of modern commercial hog raisers, makes greater demands on bodies, and nutritional deficiencies of one element or another are more likely to occur. Breeding at early ages, farrowing at shorter intervals, and producing larger litters all require more nutrition than would be needed by animals that are not worked as hard. A part of this factor can show up on the plus side for homesteaders, however. Many nutritional deficiencies are minor and take generations to show up. There is greater leeway in nutrition in a hog you feed for four or five months and then butcher than there is in breeding stock that will be around a long time and must produce healthy individuals for future generations.

Perhaps the most important cause of nutritional deficiencies in swine today is confinement housing. The pigs have no sunshine, no access to real fresh air, and can eat only what the caretaker brings them. A meticulously formulated ration is of utmost importance in this situation, but even then there is considerable doubt whether the most knowledgeable expert knows everything there is to know. Just look at what we're still learning about human nutrition—and the controversies that surround much of it!

From reading this brief outline it should be easy to see why the homestead hog raiser has so few problems with nutritional deficiencies. However, there are a few nutritional diseases that even the homesteader should be aware of.

• **Anemia.** Many diseases affect baby pigs, which the one-hog homestead will not be involved with. One of the most common, for example, is nutritional anemia, usually caused by lack of iron. (It can also be due to lack of copper, cobalt, and certain vitamins.) It is most prevalent in suckling pigs that do not have access to soil. The symptoms are loss of appetite, labored breathing, a swollen condition about the head and shoulders, emaciation, and death. Hog farmers usually inject 150 mg. of iron into the ham muscle at one to three days of age as a matter of routine. Most feeds contain adequate levels of iron for older pigs, but sow's milk is deficient in that element.

• **Hypoglycemia.** Another disease affecting baby pigs is hypoglycemia, also known as baby pig shakes. Baby pigs with this are weak, shiver, and fail to nurse. Heart action is feeble and the hair becomes rough and erect. Death usually follows in twenty-four to thirty-six hours without treatment, which consists of providing warmth and force feeding one part corn syrup with two parts water at frequent intervals. It can be prevented by good management of the gestating sows, which of course includes adequate nutrition.

• **Osteomalacia.** Lack of vitamin D, or inadequate calcium and phosphorus, or an incorrect *ratio* of calcium and phosphorus can cause osteomalacia, particularly during gestation and lactation. The ratio of calcium to phosphorus should be 1:2 to 1:1.5 for swine. Calcium is the most common deficiency in hogs because grains are deficient in calcium, and many pigs never get forages. Homestead pigs generally have no such problem. Symptoms of osteomalacia include stiffness of joints, failure to breed regularly, decreased milk production, lack of appetite, and an emaciated appearance.

• **Parakeratosis.** On the other hand too much calcium in the diet (above 0.8 percent) is no good either. It may result in parakeratosis, known by vomiting and diarrhea, reduced appetite, mangy appearance, and slower growth rate. This usually affects pigs from one to five months of age, which is the range for homestead hogs. Homestead pigs fed milk, legumes, and comfrey (which are all high in calcium) are prime candidates for parakeratosis. There might be some consolation in the realization that this is not a deficiency as such, but an overabundance, a poisoning. Furthermore, mortality isn't high. If it does occur under homestead feeding conditions, reduce the amount of forage fed and feed more grain. On a commercial basis add 0.4 pounds of zinc carbonate or 0.9 pounds of zinc sulfate heptahydrate per ton of feed. Parakeratosis isn't contagious, and the greatest loss comes from reduced gains and lower feed efficiency.

• **Salt deficiency.** Pigs need less salt than other classes of livestock, but they still need it. Salt deficient pigs may show depressed appetites, retarded growth, weight losses, and rough coats. They will also have a tremendous appetite for salt. Hogs may be fed salt free-choice, but if they have not had salt they will probably eat too much and encounter salt poisoning. That can result in death in a few hours or up to two days. The homestead treatment for salt poisoning is drinking large quantities of fresh water, although vets can administer water via a stomach tube or calcium gluconate intravenously if the animal is unable to drink. Beware of giving hogs access to brines from preserving and other homestead activities. One-half pound of salt per one hundred pounds of feed is adequate, or if fed free-choice, take it easy at first if the pigs seem salt starved. Do not leave salt where it will get wet and form brine pools.

- **Selenium poisoning.** This is a problem in certain regions where feeds are grown on soils containing selenium. The pigs' hair falls out (and in severe cases the hoofs slough off too), they go lame, and don't eat. Death is caused by starvation. The problem occurs in some parts of South Dakota, Montana, Wyoming, Nebraska, Kansas, and limited areas of the Rocky Mountains.
- **Blindness.** Vitamin A deficiency results in night blindness and, in severe cases, blindness. Lush forage, green hay, yellow corn, and whole milk are all good sources of vitamin A.
- **Rickets.** Rickets is caused by lack of vitamin D but also by a lack of calcium or phosphorus or an improper ratio of the two. It affects young pigs with bowed legs, enlarged knee and hock joints, and irregular bulges at the juncture of the ribs with the breastbone. Movement is painful, and swine frequently become paralyzed, but it's seldom fatal.
- **Iodine deficiency.** In the "goiter belt" of the Great Lakes region and in the Northwest, iodine deficiencies are possible because plants grown on soils in these regions are deficient in iodine. One symptom is pigs that are born hairless. Iodized salt is a simple preventive.
- **Fluorine poisoning.** This is another form of toxicity that can occur even on organic farms. The water in parts of Arkansas, California, South Carolina, and Texas contains excess fluorine and can result in such symptoms as abnormal teeth and bones, stiffness of joints, loss of appetite, decreased milk flow, diarrhea, and salt hunger. High fluorine phosphates used in some mineral mixtures may also be a cause. Fluorine is a cumulative poison, so its effects are more likely to be noticed in older animals.

• **Nitrite poisoning.** Nitrite, the reduced form of nitrate, falls into the same class. (This one is particularly interesting because of recent unfavorable publicity given to nitrites in bacon and other cured products. There seems to be poetic justice in the fact that the farm animals most susceptible to nitrite harm are swine!) In one study, four out of ten farm wells in one Missouri area contained more than 10 parts per million nitrate-nitrogen, the maximum permissible amount for human water supplies. Some of them were as high as 190 ppm, which approaches the lethal amount for swine under certain circumstances. Excess nitrate (or nitrite) in water supplies is a common problem over much of the north central United States. Shallow wells hold more potential danger.

Hogs can tolerate nitrate but not nitrite; however, nitrate may be converted to nitrite. In shallow wells, coliform bacteria from surface contamination can convert nitrate to nitrite. The conversion can also take place in galvanized pails or other containers because of the presence of zinc.

The main problem appears to stem from nitrite interfering with proper nutrition even though the diet is adequate. This interference results from the action of nitrite on vital enzymes and endocrine systems.

One researcher maintains that most nitrates in wells originate in organic matter on the surface: manure piles, straw and hay stacks, and other places naturally high in nitrogenous substances. This doesn't satisfactorily explain how almost half of the wells surveyed in central Missouri in 1962 were hazardous according to government standards. What about nitrogen fertilizers, the use of which has been growing at a prodigious rate?

At any rate, nitrate and nitrite are problems for some hog raisers.

In the main, nutritional deficiencies are problems for large-scale confinement hog raisers and test-tube technicians. Homesteaders who use common sense, provide a variety of feeds, and good management needn't lose any sleep over their feeding program.

Chapter 9.
HOG HEALTH

All controversies notwithstanding, there can be little or no disagreement that disease is simply the absence of health, and that health is maintained by proper diet, fresh air and sunshine, exercise, and a sanitary and comfortable environment. It might take some effort and knowledge to provide these, but not nearly as much as it takes to cure a sick pig.

In the improbable event that you encounter a sick hog, the wisest course of action is to call a vet or, at the very least, an experienced neighbor. Many swine diseases are difficult to diagnose without the knowledge and experience a one-hog homesteader might never accumulate, or have to accumulate: you could raise one pig a year for fifty years and you'll only have experience with fifty pigs. A farmer who raises ten or a hundred times that many in a single year obviously will be and must be much more knowledgeable than the homesteader, but even the farmer doesn't engage in home doctoring. The farmer's job is to keep the animals healthy. If he fails he moves over and lets the doctor practice *his* craft.

The importance of control and prevention cannot be overemphasized, especially for the homesteader. But here again, if we examine suggestions for health programs listed in various government publications, it's evident that many of the hazards to swine health simply don't exist on the homestead level. These include the recommendations that follow.

Preventative measures

• **Maintain a closed herd.** Start with stock that is free of infectious diseases and genetic defects, and don't let the animals come into contact with hogs from other herds. Isolate newly acquired swine for thirty days before allowing them to mix with your herd.

• **Make certain that trucks carrying swine are clean.** Don't let outside people or vehicles on the grounds or in buildings where swine are bred and reared.

• **Follow your veterinarian's recommendations for an immunization program** tailored to your particular needs.

• **Separate and maintain pigs by age groups,** and keep breeding stock separate from growing animals. If disease has been a problem, thoroughly clean and disinfect buildings and equipment and leave them empty for three weeks or more before moving in new stock.

• **Dispose of dead animals** through licensed rendering plants, incineration, or by burying six feet deep and covering the carcasses with lime before filling in the hole.

These suggestions aren't very applicable if you only have one or two pigs. But here are others that do affect the homesteader:

• **Start with healthy, defect-free stock.** Keep fences, buildings, and equipment in a good state of repair. Provide adequate shelter for extreme weather conditions: shade and ventilation in hot weather, proper housing and bedding in cold.

• **Provide proper nutrition;** this is the best defense against disease. Do not feed garbage unless it is thoroughly cooked. (However, as mentioned elsewhere, clean, fresh waste from your own kitchen isn't classed as garbage provided it doesn't contain raw pork or splinter-prone poultry bones.) Provide ample fresh water.

• **Protect feed and water from contamination.** This means not only contamination from pig manure, but from birds, rodents, and other animals. Don't allow poultry around your hogs.

• **Spend some time watching your pigs every day.** Be alert for signs of depression, lack of appetite or depraved appetite, diarrhea, lameness, swelling, labored breathing, unusual skin condition—in short, anything out of the ordinary.

• **When your pork is in the smokehouse, clean the pen *thoroughly.*** Remove all equipment and get every trace of organic matter from every crack and corner. It will be much easier at this point than several months later when you get another pig, and even more importantly, this will expose the quarters to the best and cheapest disinfectant known—fresh air and sunshine. If you intend to raise another pig in the same pen rather soon, it's a good idea to do the cleaning and wait at least a couple of weeks. Even then you should be aware of some of the common disinfectants used in swine buildings.

Disinfectants

One pound of lye to ten gallons of water makes a good general disinfectant. The disinfecting agent in lye is sodium hydroxide, which kills most germs and viruses. Lye is a caustic poison and should be used with care. The solution is most effective when used hot.

Sodium carbonate is mainly a cleaning agent, but it does

have some disinfecting value. Sodium orthophenylphenate, mixed at a rate of one pound to twelve gallons of water and used hot, is another recommended disinfectant.

Fumigation was a common disinfectant in the old days. Baker mentions it thusly: "Every part (of the shed) should be stopped tight, and flowers of sulphur and wood tar, in the proportion of one pound of the former and two quarts of the latter, mixed with tow, should be burned and allowed to smoke thoroughly, until the whole building is thick with smoke." Steam cleaning is more common today, usually with some disinfectant introduced into the spray of steam.

No disinfectant—even the sunshine and fresh air variety—can do a satisfactory job unless all manure and other organic matter have been completely removed in advance. In this regard, elbow grease is still the best cleanser.

Worms

Regular worming, and spraying for lice and mange, are two other routine practices on most hog farms. Hogs are more susceptible to internal parasites than any other common farm animal, with the possible exception of sheep. Internal parasites have been estimated to cause *yearly* more than $65 million in damage, and external parasites more than $3 million. Because most hog farmers worm regularly, chances are good that any weanling pig you buy will already have been wormed.

If you do decide to worm your pig, wait until it is accustomed to its new home—after ten days or so. The new pig is already under stress, and vermifuges or wormers, being poisons, also create stress. Many different general pig wormers are available. Some are administered in water, others in feed. You can purchase them from feed dealers, farm supply stores, or from a veterinarian.

However, many homesteaders do not even bother with worming. Worms have very rigid life cycles, and by breaking the cycle the parasites are controlled if not eliminated. Strict sanitation will do the trick, and that will be enhanced by the small number of pigs on the homestead as well as by the vacating of pens for long periods of time between pigs, which is common in homestead management. Add to that the somewhat mysterious effect of milk on worms in hogs. You *probably* won't have to worry about worms if you feed your pigs milk, raise only one or a few at a time and get them from a clean herd, leave your pen clean and empty for at least a couple of weeks between pig crops, and pay attention to thorough and regular cleaning.

Hogs are susceptible to many different worms, including roundworms, lungworms, nodular worms, ringworms, screwworms, stomach worms, threadworms, and whipworms. Not all wormers kill all worms, which is why it's necessary to have a vet identify the parasites present and prescribe specific remedies.

An understanding of the life cycle of these parasites will explain why control measures are important and how they work. Let's take the roundworm as an illustration. It's probably the most common, and accounts for about half of the economic loss from internal parasites. A roundworm is yellowish or pinkish, as big around as a lead pencil, and eight to twelve inches long. It lives in the small intestines where the female lays eggs that are excreted with the feces.

A small larva develops in each egg. It just sits there, waiting. If a pig happens to swallow it, the larva emerges from its shell, bores through the wall of the intestine, and enters the bloodstream, ending up in the lungs after a brief stopover in the liver.

In the lungs it breaks out of the capillaries, enters the windpipe, and migrates to the throat where it is swallowed and carried to the intestines to develop into a sexually mature worm. It lays eggs that are excreted in the feces—but we don't have to go into that again because that's why it's called a cycle.

It's obvious that if the cycle can be broken at any point the worms would be eliminated. There are two places to do this.

Vermifuges kill the worms already in the pig. From that standpoint they are the best solution. Atgard V, the trade name for dichlorvos, is a prescription drug available through veterinarians which kills worms of many kinds, including roundworms. There are several piperazine compounds which kill roundworms but have little effect on others. Use any of these according to the manufacturer's directions.

The other point at which the cycle may be broken is when the eggs are in the feces. Remove the manure and the eggs for composting, and voilà, no more worms. If you start out with worm-free pigs, sanitation will be the only worm prevention you'll need, and of course it's important when poisons are used.

Another worm with a slightly different life cycle is of interest to homesteaders using pasture. That's the lungworm. Swine first acquire it by eating infested earthworms. How do the earthworms get infested? By feeding on swine manure that is infested with the eggs of the lungworm that lives in the swine. The cycle, again. This cycle demonstrates the need for pasture rotation.

For at least part of the cycle, parasites can only exist in the bodies of their hosts. That means they start, and end, with the pigs. Buying clean stock cannot be overemphasized. Your chances for raising worm-free pigs are greatly enhanced if the premises of the seller indicate that sanitation is an important part of his management.

Lice and mites

Hogs are affected by one species of lice and by two species of mites that cause mange.

The hog louse is one of the largest species of lice, and since some hog lice are as much as one-quarter inch long, they can easily be seen. Since their entire life cycle is spent on the hog, they are transferred from pig to pig. Lice suck blood and should be eliminated with insecticides. Mites are much smaller than lice, the most common (Sarcoptic) being only one-fiftieth of an inch long. Adult females burrow into the pig's skin and lay eggs in the tunnels. Such chemicals as lindane, malathion, and toxaphene control both lice and mites on hogs. Follow the manufacturer's directions.

Major diseases

The importance of working with a qualified veterinarian in the event of disease has already been stated. However, the following brief outlines of some major swine diseases should help make the homesteader or small producer more aware of some of the possibilities.

• **Anthrax.** Anthrax is usually fatal, killing by suffocation and blood poisoning. Swine with anthrax usually have swollen throats, high temperatures, and pass bloodstained feces. The disease is identified under the microscope by the presence of *Bacillus anthracis* in the blood. It can affect humans. The anthrax bacillus can survive in a spore stage for years, generally on pasture. This disease can be prevented by immunization, but treatment has not been satisfactory.

• **Atrophic rhinitis.** This is probably a disease complex rather than a single disease, and some evidence indicates it can also be a nutritional factor linked to a calcium-phosphorus imbalance or deficiency. Persistent sneezing in young pigs is the first sign, followed by a wrinkling, thickening and bulging of the snout. At eight to sixteen weeks of age the snout and face may twist to one side in a rather grotesque fashion. It is often accompanied by nosebleeds. Affected pigs make slow and expensive gains, but if death occurs it's usually due to pneumonia. The best preventive measure lies in selecting breeding stock that is known not to be a carrier. Affected pigs can be put on a creep feed containing a hundred grams of sulfamethazine per ton of ration.

• **Brucellosis.** Also known as infectious abortion, brucellosis causes losses of unborn pigs, but perhaps the greater danger is that it can be transmitted to man. Bang's disease in cattle, Malta fever in goats, and Traum's disease in hogs are all forms of brucellosis, and any can be transmitted to man where it is known as undulant fever, Malta fever, or brucellosis. The signs in swine are indefinite as not all abortions are caused by brucellosis, and sows infected with brucellosis do not necessarily abort. The brucella organism is acquired through the mouth by licking infected animals or eating contaminated feed or water. Boars can transmit the disease when breeding. Control is a management task for breeders; they should make sure that their sows and boars are free of the disease. Control is the only answer, as there are no known cures.

• **Hog cholera.** This is a highly contagious disease affecting only swine, which wiped out entire herds in many parts of this country earlier in this century. Despite widespread control measures, the disease has not been eradicated. It strikes quickly, accompanied by fever, loss of appetite, weakness, a purplish coloring of the underside, coughing, discharge from the eyes, chilling, and constipation alternating with diarrhea. On the other hand, some pigs die without showing any symptoms at all. And to further obfuscate matters, the disease is often confused with erysipelas.

The National Hog Cholera Eradication Program, enacted in 1961, provides for a federal-state program to wipe out the disease through such measures as vaccination, eliminating or cooking garbage fed to swine, and eliminating all possible sources of the virus by identifying and destroying any hogs with cholera. Hog cholera has been wiped out in Canada, and that goal seems close in the United States.

• **Swine dysentery.** This acute infectious disease is usually associated with animals that pass through central markets. The main sign is profuse and bloody diarrhea. Sanitation is the key to control, and that includes buying clean animals. Be wary of auction barns and similar markets where large numbers of swine pass through. Some animals can be carriers—they appear healthy but can infect others. Veterinarians can reduce death losses with antibiotics, arsenicals, and nitrofurans, but that's small consolation to the homesteader whose one or two pigs are already dead.

• **Enteritis.** Enteritis is a general term covering several diseases which produce an inflammation of the intestines. It can be associated with poor sanitation, B-complex vitamin deficiencies, or internal parasites. Good management, which includes proper nutrition and sanitation and quarantine of new animals, is a preventive measure.

• **Enterotoxemia.** Enterotoxemia or edema disease strikes pigs six to sixteen weeks of age, and is usually fatal. Death can come in a matter of hours. The disease usually affects the best and biggest piglets, with fever, swollen eyelids, and often constipation. Paralysis sets in. The cause isn't known, and there is no satisfactory treatment.

• **Erysipelas.** This is another of the diseases that can be transmitted to man, although in humans it is called erysipeloid since human erysipelas is something else. In swine it is often confused with cholera. The causative bacteria can propagate both in the animals and in the soil, making prevention difficult. Control is definitely a job for a vet.

• **Foot-and-mouth disease.** While it has not been reported in this country since 1929, foot-and-mouth disease does exist in other areas, and strict measures are undertaken to prevent its recurrence here. In the outbreaks which have occurred in the United States, every affected and every exposed animal was promptly slaughtered and destroyed. No hogs or uncooked pork products are allowed to enter the United States from any country where foot-and-mouth disease exists. There's no need for homesteaders to worry about this one at all.

• **Hog flu.** This flu is more serious than the kind humans get, although they resemble each other. It's due to a combination of bacteria and a virus, but in a really far-out web of ecology. The bacteria alone don't cause the disease. The virus is found in the lungworm. From the life cycle of the lungworm you'll recall that the intermediate host is the earthworm. The pig must eat an earthworm which is infected with a lungworm which is infected with the virus, and that's how the virus gets together with the bacteria to cause the influenza. There is no treatment, but some people believe death losses are due to secondary infections that can be treated with sulfa drugs or antibiotics.

• **Leptospirosis.** This disease affects swine, cattle, dogs, and man. Swine losses take place when pigs are aborted or born weak, and also in the general unthriftiness of infected market hogs. The infection is usually confirmed only after laboratory tests are undertaken after abortions occur. Blood tests and vaccinations are the usual preventive measures, along with ordinary management practices.

Fairly good results can be had with prompt antibiotic treatment.

• **Mastitis-metritis-agalactia complex.** Known also as the MMA complex, it is a disease of sows which often results in the loss of baby pigs. It is an inflammation of the mammary glands and uterus, and/or a failure to secrete milk. The disease is complex, and the actual cause is unknown. Good management practices seem to be the best general preventive measure.

• **Mycoplasma pneumonia.** This is said to be the most important pig disease in terms of economic loss. Feed efficiency of affected animals can be lowered by 25 percent. Coughing and diarrhea are the chief signs. Pigs may eat well, but do not gain weight. It has been estimated that one sow in four is a carrier, which means the disease is continued in infected herds by contact between sows and piglets.

• **Pneumonia.** Pneumonia is an inflammation of the lungs that affects all animals. Untreated cases have a 50 to 75 percent mortality. The signs are chills, high temperature, shallow breathing and gasping for breath, coughing, and a discharge from the nostrils and eyes. Pneumonia can be caused by many things, including irritation of the lungs from such sources as drenches given by unskilled persons, lungworms, viruses, and many microorganisms. Preventive measures, as usual, center on general good management. Treatment should be left to a vet.

• **Scours.** Scours is often fatal. It affects baby pigs with creamy yellow or grayish green, watery feces. The pigs may be dehydrated. Causes may include dampness and chilling and improper rations. You should maintain sanitation and provide sufficient bedding. Oral treatment with neomycin or a nitrofuran drug is often effective.

• **Shipping fever.** This can be a serious problem with any livestock that is moved, especially under other stress-promoting conditions. The signs are high temperature, difficult breathing, coughing, discharge from the eyes and nose, and sometimes diarrhea. Sometimes animals die of shipping fever without showing any signs at all. Because the predisposing factors are overcrowding, mishandling, and shipping that covers several days (especially in winter in unheated vehicles), homesteaders rarely see this disease.

• **Swine pox.** While a relatively minor disease in that death losses are rare, this disease is common, especially in the Midwest. Small red spots are likely to appear on the ears, neck, and inside of the thighs, and they become roughly dime-sized and develop a hard nodule in the center. These turn into small blisters which contain a clear fluid that later becomes puslike. The blisters then dry up, leaving brown scabs which fall off. The disease is transmitted by insects, primarily lice, so lice control is the best preventative. There is no treatment as such.

• **Transmissible gastroenteritis.** TGE, as it is known, is a fairly common hog ailment that involves inflammation of the stomach and intestine. It is accompanied by scouring and often vomiting, and it causes high mortality among pigs less than a week old. It is thought to be transmitted by a virus. There is no effective treatment, but the best prevention, once again, is good general management.

• **Tuberculosis.** Tuberculosis comes in three varieties: human, bovine, and avian. Pigs can get all three. This is usually a disease of the lungs and lymph nodes, although it can also affect the udders of cows; the liver, spleen, and intestines of poultry; and the stomachs of hogs, which usually get the infection by eating infected material. The disease is becoming rare through stringent eradication measures inaugurated in 1917. Tubercular animals must be disposed of under that program. Strict sanitation and pasture rotation are important preventative measures, and chickens should not be allowed to mingle with swine.

The normal rectal temperature of pigs is 102.6° F., ranging from 102° to 103.6°. Normal pulse rate is sixty to eighty per minute. Normal respiration is eight to eighteen per minute.

These brief discussions of swine health problems demonstrate several very important points. First, disease is much more easily prevented than cured. Diseases require the services of a trained veterinarian. And perhaps most importantly, many of the factors that predispose animals to disease simply don't exist on the homestead, where animal numbers are small, nutrition is adequate, and general daily care and sanitation are excellent.

Chapter 10.
BUTCHERING

There is more than one way to skin a cat, and there are a couple of ways of converting your pig into pork, too. The following procedures are only recommendations for those who don't have experience or ideas of their own.

I say this right at the beginning, because every item I've written about butchering hogs has resulted in letters from readers who do things differently. I've tried some of their suggestions and liked some of them; some of them I didn't like, but if they work for others that's their business.

Actually, the controversy starts at the very beginning. Some people say it's cheaper, and certainly easier, to have an abattoir do the butchering. My own experience indicates otherwise. Of course there are finagle factors involved: some butchers charge more than others; some people have an easier time transporting a hog than others; a homesteader with some butchering and meat-cutting experience will have a much easier time than one who faints at the sight of blood or guts; and some people might think their time can be spent more profitably on other tasks. The homesteader interested in reducing the grocery bill will have a different outlook than the one who raises the pig just for kicks.

Here's how I would suggest approaching this problem. First, find out how much it would cost to have a custom slaughterhouse do the job. Figure that into the cost of the pork that will reach your table and decide if the price is worth it.

Do you have a means of moving the hog—a pickup truck with side racks and a loading ramp of some sort? If not, what will it cost to have someone else transport the animal? Are you physically and psychologically capable of killing a pig and wrestling with a two-hundred-pound-plus carcass? How important is meat quality (which will probably deteriorate by transporting the animal; see the chapter on quality) in your scheme of things? All these factors, and more, enter into your final decision.

No doubt an ideal middle ground is to engage the help of an experienced friend or neighbor, especially if he has some of the tools you'll be needing. Help will be necessary anyway, and if it's experienced help, you'll learn a lot more than you will by reading a book or by groping around by yourself. Of course, it would also be nice not to have to put out any cash for some of the rather specialized tools you'll be needing. Lots of people will be more than glad to help out for a couple of packages of roasts or chops.

One quite common error beginning pig raisers make is letting the animal put on too much weight. Bigger is not better! Weight gains above 220 pounds or so are more expensive than gains at lighter weights, because it takes more pounds of feed to produce a pound of pork above that range. Also, most of today's families prefer smaller cuts, which come from smaller hogs. And finally, pigs that weigh much more than 220 pounds will have more fat, which makes curing more difficult and decreases overall quality.

The necessary equipment

If the pig is to be scraped (and we'll come to this in a moment), a scalding vat is required. A fifty-gallon drum works well and is readily available in most places. Other people have used stock tanks or even old bathtubs. Anything watertight that will hold the pig will work.

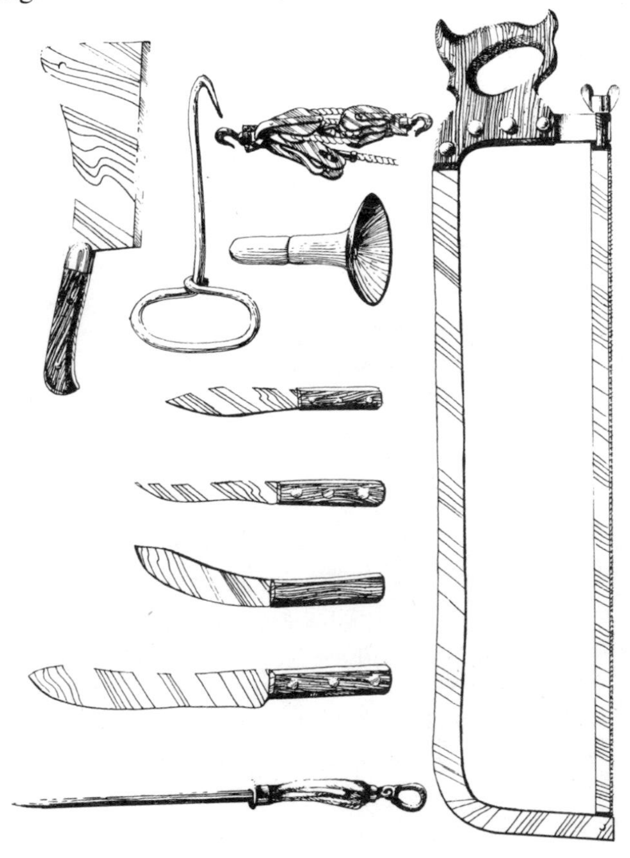

Butchering tools.

You'll need a place to hang the carcass while it's chilling—a place where it will be safe from dogs, dust, insects, and other hazards. At the right time of year in the right kind of weather, a tree might serve the purpose. Or you might use the rafters in a barn, shed, or garage.

A bell hog scraper is a must, in my opinion, although some people claim they can use a dull butcher knife successfully. Hog scrapers are available from the Countryside General Store, Waterloo, Wisconsin 53594.

A saw of some sort is a necessity. If you don't have access to a real meat saw, the handsaw in your workshop will do an admirable job, and no one raving over the pork chops and applesauce will ever know the difference.

A variety of sharp knives, and the stone and steel to keep them sharp, will make the job much easier. You'll need some rope, and a light block and tackle (if you don't have enough help to hoist a two-hundred-pound carcass).

A meat pump will be handy for curing hams, especially if they're large, but don't let the lack of this item keep you from butchering a hog. A cleaver might have some application but it isn't a necessity. Use it sparingly anyway to avoid bone chips. The saw is better.

You can make the job easier and neater by keeping the animal off feed for twenty-four hours before butchering. Provide water though.

The slaughter

There are several ways to do the actual killing, each with its own proponents. There is little doubt in my mind that the most humane method, and also the one most conducive to quality pork, is to merely cut the jugular, without shooting or stunning with a sledge. This operation can be conducted with the hog in a confined area, but it's easier and better if a rope is placed around a hind leg above the hock and the animal hoisted with its head just off the ground. Another method is to throw the hog on its back by grasping underneath for a leg on the opposite side and throwing it off balance.

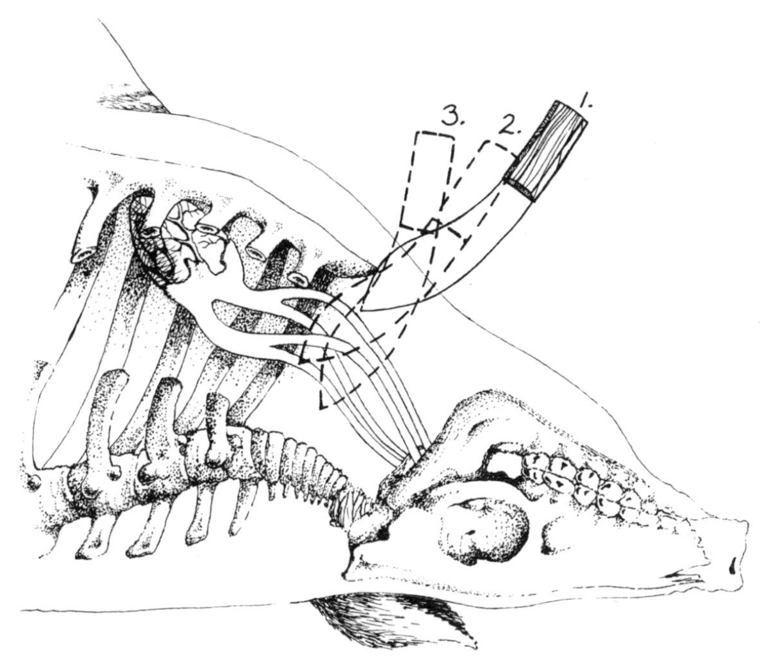

Procedure for sticking, showing sequence of knife movements.

Make an incision from the point of the breastbone, two to four inches forward down the middle of the neck. Then insert the knife in this incision at an angle of forty-five degrees and force it down and back to a point about six inches below the front of the breastbone. Give the knife a slight twist before withdrawing it. This will sever the large branching vein and the artery which lie beneath the point of the breastbone, hopefully without puncturing the chest cavity, which would permit blood to settle in that area.

Gruesome as this might sound to the neophyte, an experienced helper can have the job done in less time than it will take you to check the temperature of the scalding water. And at one point in his career that experienced pig sticker didn't know any more about it than you do.

When bleeding is complete and the hog is dead and the water is 145° F., dunk the carcass. If you're using the fifty-gallon drum, it should be leaned up against a scraping table (our picnic table is the right height) at about a forty-five-degree angle so that you're really dragging the carcass more than lifting it. It should take two to three minutes of scalding at this temperature to sufficiently loosen the hair. Some people claim hotter water works better for them, and some like to toss a handful of wood ashes in the water to help loosen the hair.

We usually swish the pig around in the water and test the hair by twisting some off with our fingers. When it comes out easily, remove the carcass and start scraping the hair off as rapidly as possible.

The hog scraper is used by applying a steady pressure in a circular motion. Tough spots require tilting the scraper on an edge and pulling forward. Because scraping the feet and head is hardest, some people do these first. It will probably be necessary to dunk the carcass back in the hot water as the scraping gets more difficult. Some touching up with a knife might be required. After all the hair and scurf is removed, even a black pig will be white. While the carcass is still hot, pull off the dewclaws and toes with a hook. (A bale hook from the hay mow works.) When the job is complete, rinse off the carcass and hang it with the head about six inches off the ground.

Alternate methods of preparing the carcass as reported by homesteaders include laying gunny sacks soaked in boiling water on the animal; blistering the carcass with a blowtorch and then scraping it; and skinning it. None of these have worked well for me, and the only comment I can make is that if you decide to skin your hog, definitely plan on using a skinning knife, which has a special curve to the blade and makes the job considerably easier. There will still be a tremendous waste of lard.

When the carcass is cleaned, rinsed with cold water, and hanging, remove the head. Cut just above the ears at the first point of the backbone and across the back of the neck. Sever the gullet and windpipe. Continue the cut around the ears to the eyes and to the point of the jawbone, which will leave the jowls on the carcass.

To open the carcass, cut down between the hams to the incision made when sticking the pig. The breastbone will have to be sawed in two, and the aitchbone will have to be split. (You'll recognize this bone when you come to it.) This can best be accomplished by placing a stout knife against the bone and striking the butt of the handle sharply with the palm of your hand. Loosen the bung by cutting around it, and pull it down.

By then you should be pretty well into it. You'll be able to remove the entrails with a little studied grasping and pulling, and perhaps a little cutting, especially around the diaphragm. This is the part for which keeping the animal off feed beforehand helps, because empty intestines and stomachs are easier to work with.

Cut the liver from the offal, and carefully remove the gall bladder. Hang the liver on a peg inserted through the thick end, and split the thin end to facilitate drainage. Cut off the heart, wash it, and hang it by the pointed end to facilitate drainage. If the intestines are to be used for sausage casings, have someone turn them inside out, wash them, scrape them with a blunt stick, and soak them in a weak solution of lime water (see your druggist) for twelve hours. Lacking lime water, use one tablespoon of baking soda in two gallons of water.

Meanwhile, the carcass has been washed once again and split down the middle of the backbone with the saw. The leaf lard (it will be apparent when you reach this point) is removed by pulling.

The next step is cooling—one of the real hangups of home butchering. The ideal temperature is 34° to 40° F. If the carcass freezes before all of the animal heat is out of it, meat souring may result. On the other hand, temperatures that are too high permit the growth of unfriendly bacteria. Lacking a walk-in cooler, you must rely on nature; keep a weather eye out when planning to butcher.

The meat should cool for twenty-four hours. Then the cutting can begin. Technically, the temperature at the center of the hams should be 33° to 35° F.

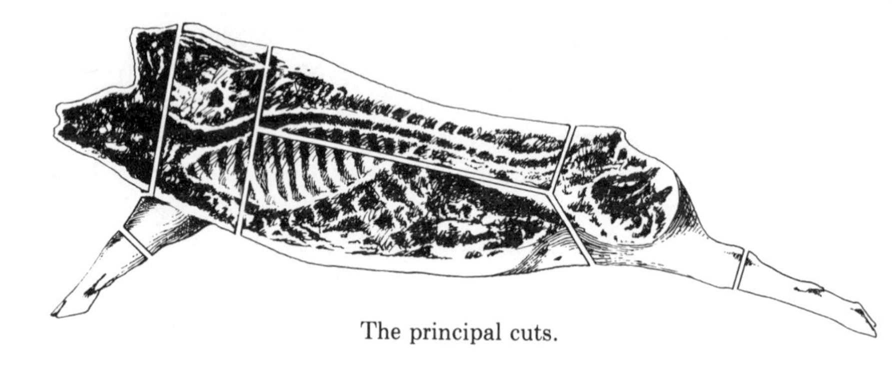

The principal cuts.

Cutting up the carcass

Some people seem to have a knack for this. They can operate on a side of pork and come up with rolled roasts and tenderloins and blade chops that could be used in TV supermarket ads. Others of us figure meat is meat, and more or less hack away. It tastes the same either way, and you really can't go wrong. But here are some guidelines.

The principle cuts of pork are the ham, loin, bacon, shoulder, and jowl. Everything else is classed as trimmings, along with loose pieces trimmed off the main cuts.

You'll be working on one-half at a time, probably on a paper-covered table or bench. The first cut is made at the shoulder, sawing between the third and fourth ribs at right angles to the back (A). Once through the bone, a large sharp knife works better than a saw. With the skin-side up, cut off the jowl at the point where the backbone ends (B). Trim some of the cheek meat and flatten it with the broad side of a cleaver or hatchet, then trim it square with a knife to make a "bacon square" that can be cured like bacon or used as seasoning meat (C).

Remove the neck bone from the shoulder, and trim off the meat. Cut off the shank above the knee joint (D). The shoulder can then be cured as is, or divided between the smallest part of the blade bone to produce a picnic shoulder and a butt. The fat on the top of the butt can be trimmed off and either cured for seasoning or used as lard. The lean por-

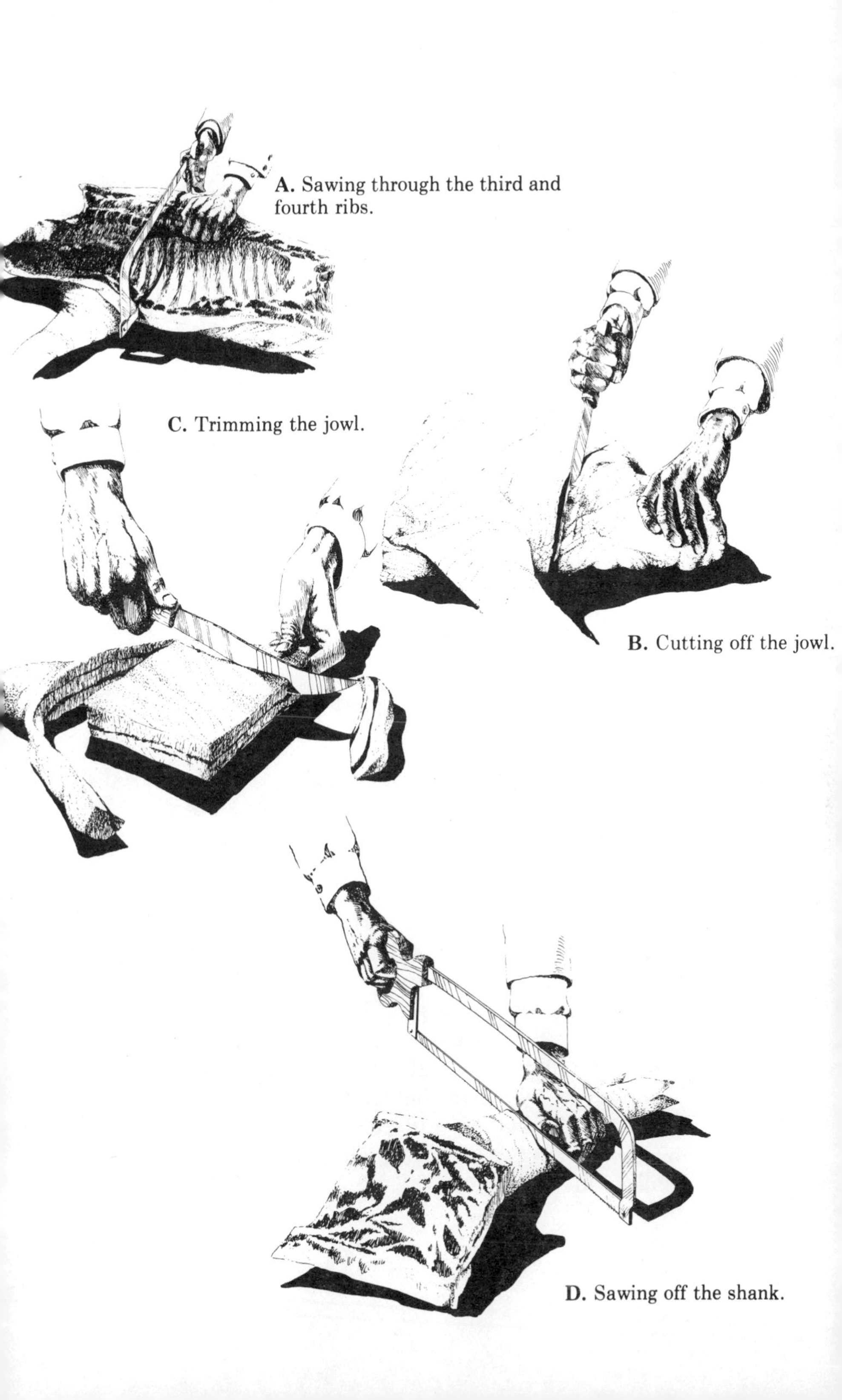

A. Sawing through the third and fourth ribs.

B. Cutting off the jowl.

C. Trimming the jowl.

D. Sawing off the shank.

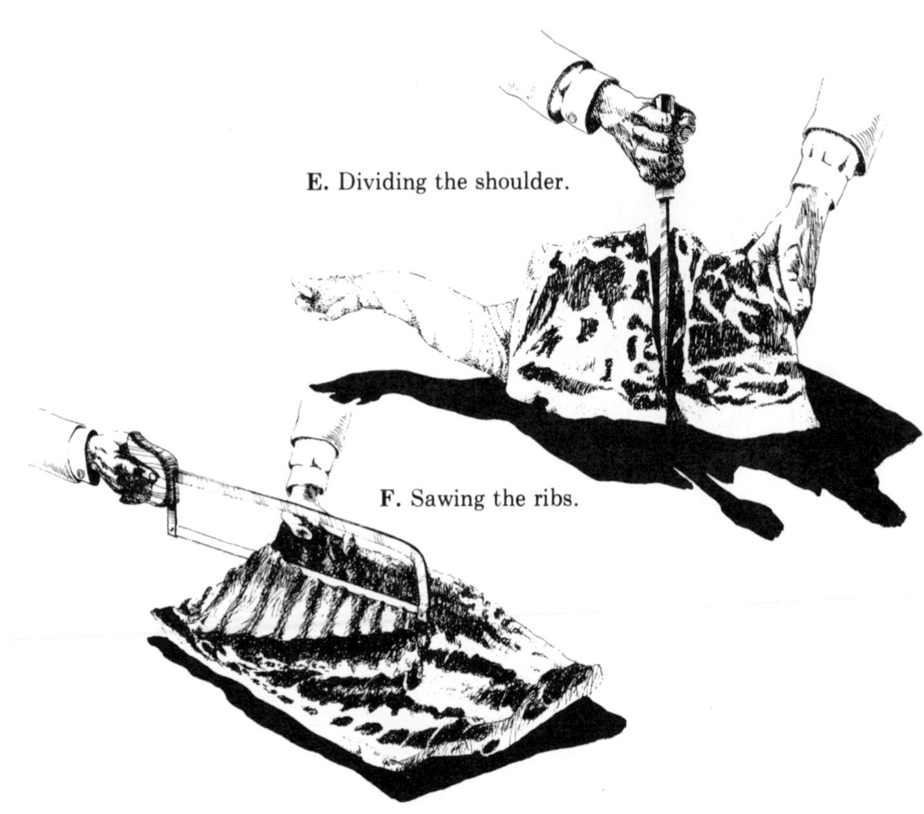

E. Dividing the shoulder.

F. Sawing the ribs.

tion is the Boston butt, which can be cured or used for sausage. The picnic shoulder will look like a small ham.

Remove the ham by sawing on a line at right angles to the hind shank and at a point a couple of inches in front of the aitchbone. Finish the cut with a knife (E).

Remove the tail bone by slipping the knife under it, cutting along it with the knife flat. Trim off all loose pieces and corners for sausage, because if left on the ham they will dry up in the cure and be useless. The hams can be skinned, but they do not keep as well.

Saw the shank off just below the button of the hock. Separate the loin from the belly by sawing across the ribs about one-third of the way from the top of the backbone to the bottom of the belly (F). The tenderloin should be included with the loin.

Lay the belly on the table skin-side up, smooth out the wrinkles as best you can, and give it a few good blows with the broad side of a cleaver to help loosen the spareribs. Then turn it over and trim out the ribs. Start by loosening the neck bone at the top of the ribs. Be sure to keep the knife as flat as possible to avoid gouging the bacon, because the flaps will be waste after the bacon is cured. Trim as close to the ribs as you can.

Choice bacon is trimmed and squared. Start on the lower edge by cutting it straight and trimming out the mammary glands. Then trim the top parallel to the bottom, and square off both ends. Trimmings are used for sausage or lard.

You'll find a small lean muscle underneath the backbone toward the rear of the loin. That's the tenderloin, which some consider the filet mignon of pork. Trim it from the loin, and make sure it doesn't end up in the kettle for sausage meat! Remove the backfat from the loin by cutting along its length, leaving about one-quarter inch of fat on the loin. Backfat can be used for lard, or cured and used for seasoning baked beans and similar dishes.

The only practical way for the home butcher to cut pork chops (lacking a power meat saw) is by cutting between the ribs. This makes a sinfully thick—but positively scrumptious—chop.

As has been mentioned, there really is no way to go wrong when cutting pork. We might not all make a cut in the same way, but the pork will be edible nevertheless. And even just a little experience will be a great teacher. With that in mind, start on the second half just like you did the first, and see if there isn't some improvement already. Much of the cutting depends on personal preferences. Some people will want larger cuts than others want, some prefer roasts to chops, and some will want stew meat more than sausage. Similarly, some will be more interested in curing certain cuts of pork while others will prefer them fresh.

Curing

Be sure not to begin curing any meat until it is cooled to at least 40° F. Meat that is cut up and put in cure before the body heat is out becomes a prime candidate for spoilage. Of course, meat spoilage can occur for additional reasons: excess blood can cause spoilage to start, which is one reason proper bleeding is important; hogs that are excited or heated before butchering can result in a spoilage problem; and meat should not be frozen while chilling or curing.

There are many different curing methods, but they can basically be classified as either dry curing or brine curing. With either method, "pumping" the larger cuts is a good idea in order to let the curing work from the inside out as well as from the outside in. The operation requires a meat pump, a tool that looks very much like a giant hypodermic. Two pounds of cure are mixed with three quarts of water. This mixture is injected into larger pieces of meat, especially hams, because bone souring is a real threat to the outcome of your curing. Lacking a meat pump, it might be wise to consider boning the hams or making larger cuts into smaller pieces.

One sweet-pickle brine that has given good results contains 8 pounds of salt, 2 pounds of sugar, and 4½ gallons of water. (The recipe really calls for 2 ounces of saltpeter, too, but people who are concerned about nitrites will question that. For the homesteader who will put the resulting hams and bacons in the freezer anyway, the most noticeable difference the lack of saltpeter will make will be a somewhat less appetizing color. See Appendix D for more information.)

Put the meat to be cured into a stone crock, wooden barrel, or plastic garbage can (a new and clean one, obviously). Naturally you won't use a metal container for brine. Pour the brine over the meat, and weight it down to keep it from floating. It takes four days per pound to cure meat: a fifteen-pound ham would take sixty days; bacon and other small cuts require proportionately less time.

Overhaul the meat on the seventh, fourteenth, and

twenty-eighth days—this means taking the meat out and repacking it so the brine is evenly distributed and the cure will be even. If the brine gets very slimy, discard it, rinse off the meat, and make a new batch of brine. Make it proportionately weaker or the meat will be too salty. For example, if the meat has cured fourteen days, make the new brine only half as strong.

The dry cure is a little more work, but in my opinion it's worth it. Use five pounds of brown sugar; five pounds of noniodized (preferably) salt; two ounces of black pepper; two ounces of cayenne pepper; and two ounces of saltpeter (again, optional). Combine all the ingredients, and then rub the mix all over the meat. (It might be a good idea to wear rubber gloves during this part.) Work it in well, especially around the bones. Let the meat set overnight, away from flies, cats, and dogs. A lot of liquid will drain from the meat, so make certain the work surface is well padded and that the area will not be damaged by the seepage. The next day rub the pieces with more curing mix, and turn them upside down. Repeat this for seven days.

According to the original directions, the next step involves storing the meat in an insect- and rodent-proof box, with alternate layers of wheat bran and meat. If the bran imparts any flavor to the meat, I can't discern it, and we have a deuce of a time getting the bran off the finished product. I suspect the bran is used just to soak up the juice. Other people have used oats instead of bran, but I've taken to just packing the meat, well covered with the curing mix, in a box or crock. A box or wooden barrel is best, and it should have several holes bored in the bottom to let the bloody water drain out. Leave it (undisturbed, the recipe says) for six weeks.

It's been claimed that for the best flavor, bacon should be "seasoned out" for two weeks after going through the cure, and hams should age two months or more. A properly cured ham will keep for a year or more in a cool, dark, well-ventilated place, if saltpeter was used in the cure.

Chapter 11.
USING PORK

Lard

While the meat is being smoked with coals of hickory or apple, there is lard to be rendered in the kitchen. The best way to render lard is to grind the fat, put it in a large pan, and put the pan in the oven. When the fat is melted out, put it in a lard press to extract the last of it.

Lacking a grinder, cut the fat into pieces as small as possible. Lard can also be rendered in a kettle on the stove top, but the fumes are less noxious in the oven. A lard press is something of a luxury, costing well over $100. If you don't have one already, the amount of lard you'll lose will hardly be enough to make such an investment worthwhile.

To clarify the lard and make it whiter, drop in a few slices of peeled raw potato until they soak up the impurities. The cracklings—the bits of meat and rendered out fat that remain after pressing out the lard—are a gourmet item on some homesteads.

Sausage

Sausage making is an art in itself, and it's doubtful whether any homesteader could live long enough or eat enough to try all the possibilities. But since pork is a basic ingredient in sausage, and since recipes are fairly hard to come by, here are some suggestions that will get you through at least a couple of butchering sessions.

A sausage grinder is indispensable. An ordinary food chopper just isn't sturdy enough.

Basically, sausage is ground, seasoned meat. But it can be blended in innumerable ways, cooked, cured, air-dried, and smoked, so that the possibilities are endless. One of the basic sausages on the hog-raising homestead is plain old pork sausage, or breakfast sausage. And here is one of the standard recipes for this sausage.

Grind 5 pounds of pork trimmings, using one-third fat and two-thirds lean. Grinding twice produces a finer product. Season with 5 teaspoons of salt (sea salt will work in any sausage formula), 4 teaspoons of ground sage, 2 teaspoons of ground pepper, 1 teaspoon of sugar, and ½ teaspoon of ground cloves. This sausage can be wrapped and frozen in family-size portions. Just make little patties to fry it.

When using a grinder, run the meat through a coarse plate first and then through a one-eighth-inch plate. Make sure the blade and edges of the holes in the chopper plate are sharp, or the auger will force the meat through in strings; this isn't always noticeable while you're grinding, but it will be when you're chewing and then of course it's too late.

Sausage casings can be bought from many butcher supply houses, or perhaps a local butcher will sell you some. They are also available by mail. You can even make your own casings. Remove all fat and membranes from the outside of the intestines, and turn them inside out. Clean the inside with borax water, or bleach the intestines by soaking for twenty-four hours in a solution of one ounce of lime chloride in one gallon of water. Scrape away the slime and inner lining to get them as thin and transparent as possible.

For some sausages (such as salami) a larger casing is required. Lacking beef intestines, a suitable casing can be made from muslin. Stitch strips of muslin to form bags about 2½ inches in diameter and 15 inches long. These should be dipped in water and wrung out before being used.

The lard press mentioned earlier doubles as a sausage stuffer, and some grinders also come with stuffing attachments. You can also purchase a hand sausage stuffer, which looks something like an overgrown funnel. Which you use depends on your finances, and to some degree no doubt on how much sausage you intend to make over the years.

Other good recipes for sausage follow.

Paul's Liver Sausage

2½ lb. pork liver
10 lb. lean pork trimmings
2½ lb. pork fat
½ tsp. ground thyme
1 tsp. ground allspice
½ tsp. red pepper
1 tsp. black pepper
½ tsp. nutmeg
⅓ cup salt
1 bay leaf
2 cups cornmeal
⅓ cup brandy

Simmer the trimmings with the bay leaf for two hours. Reserve the liquid. Cut slits one-half inch apart in the liver, and simmer for ten minutes. Grind the trimmings and the liver. Mix the spices with the trimmings, liver, cornmeal, and brandy. Add enough liquid to form a heavy paste, and stuff into casings. Simmer the stuffed sausages for ten to thirty minutes, then immediately chill in ice water until cooled.

Hard Salami

40 lb. chuck beef
40 lb. pork jowls (glands trimmed) and pork shoulder
20 lb. regular pork trimmings including some hard backfat
3 lb. 8 oz. salt (noniodized)
1 lb. 8 oz. sugar—turbinado or white (Honey takes a lot of adjusting.)
3 oz. white pepper (Black pepper tends to discolor in splotches during the cure.)
1 oz. whole white pepper
½ oz. saltpeter (This is a must in sausage of this type to prevent botulism.)
⅜ oz. garlic powder or one quart of crushed garlic cloves

Grind the beef through a one-eighth-inch plate and the pork through a one-quarter-inch plate. Mix it all up, making sure the lean and fat are well distributed. Spread this on curing trays, no more than three inches thick. Trays can be homemade from hardwood, or you could use porcelainized steel. Anything else will leave a taste. If you must use another material, cover it first with waxed paper, although this surface will be difficult to work with later during the mixing.

Sprinkle the spices and curing formula over the meat, and let the trays set for four days at 38° to 42° F. Remix each tray at least three times a day for the first two days and once a day thereafter. Then stuff the meat into casings, pressing it in tight with your thumbs. Beef middles or pure lard-dipped muslin make the best casings for this type of sausage. Lightly salt the outside of the casings after stuffing.

The secret to hard salami is in the drying. Dry at 40° F. with 60 percent relative humidity. Mold indicates too much humidity; should it develop, wipe each sausage with olive oil. Drying takes six to eight weeks.

Genoa Style Salami

This is a slightly softer sausage. Use the formula given for hard salami but add eight ounces of burgundy to the spice cure. Hang first at 65° to 70° F. and 60 percent relative humidity for six to eight days, and then at 48° to 53° with 60 percent relative humidity for ninety days.

Salami #2

10 lb. pork
10 lb. lean beef
  (or other meat,
  such as venison)
1½ lb. onions
1 tbs. garlic powder
8 oz. salt
4 tsp. black pepper
4 tsp. white pepper
40 oz. dry red wine (optional)

Dice the onions, cut the meat into chunks, and mix everything except the wine. Grind it twice. Add the wine and mix thoroughly. Refrigerate it for two days and stuff it into large casings. Refrigerate for two more days, and then smoke at 85° to 90° F. until it is a rich brown color. Freeze this one if you intend to keep it for long periods, and cook before eating.

Pepperoni

7 lb. pork
3 lb. lean beef
9 tbs. salt
1 tbs. sugar
1 tsp. saltpeter
1 tbs. cayenne
3 tbs. paprika
½ tbs. anise seed
1 tsp. garlic powder

Grind the meat and knead in the spices, keeping the meat as close to 38° F. as possible. Cure in the refrigerator (or elsewhere at 38° F.) for twenty-four hours. Mix again and stuff into casings. Hang at 48° F. for two months.

Blood Sausage

If you have saved the blood, here is Morton's recipe for blood sausage:

3 gal. hogs' blood	½ lb. Tender-Quick pickle
7 lb. beef hearts and trimmings	1 oz. black pepper
2½ lb. pork fat	onions, mace (optional)

Stir the blood constantly while it is being collected, to remove the stringy fibers. Cook the beef and pork together for about ½ hour, and then put the beef through a one-eighth-inch grinding plate. Cut the pork into small pieces and mix with the ground beef. Then stir in four ounces, or half, of the Tender-Quick (available from Morton Salt Company), stir the meat into the cold blood, and add the rest of the Tender-Quick and spices. If this mixture is not thick enough for stuffing, add enough finely ground cornmeal to give it the consistency of thick mush. Stuff it into beef casings, and cook it at 160° F. for 1½ hours or until a sharp pin can be run into the center of the sausage and be withdrawn without being followed by blood.

Lay the sausages on a table for a half-hour, and then turn them over and let them set for another half-hour. They are then ready to use, but they will keep better after smoking for about eight hours.

Variety meats

Because pork is such an accepted feature on American menus, there probably isn't much point in dwelling on cooking it. You're certain to have your own favorite recipes already.

However, a pig is not all pork chops—or bacon or ham. If you've been buying your meat at the supermarket, chances are you've passed up such items as head cheese, tongue, scrapple, pigs' feet, and heart. But these are all part of the legacy of the homestead pork producer. You paid for them in feed and labor, and it will behoove you to learn to use them.

These variety meats can be wrapped and frozen like other fresh pork for later use. Or, they can be stored in a crock using Morton Salt's Tender-Quick pickle mixed at a rate of two pounds per gallon of water. The water should be boiled and allowed to cool before mixing the pickle. Simply place the meat in a stone crock, and pour the pickle over it. When you want anything from the crock, just take it out and rinse it in fresh water. Cook it as you normally would.

Tongue

Here's a good, basic recipe for tongue.

Place a tongue, two medium-sized onions, a carrot, several ribs of celery, and some parsley in a kettle. Just barely cover with boiling water, and add a teaspoon of salt and eight peppercorns. Simmer until tender (about three hours). Drain the tongue, skin it, and serve it. This is good with a mustard or horseradish sauce.

Or, if you can take the recipe a step further and slice it and bake it in a sauce, your family won't even know what they're eating. For this sauce, melt 2½ tablespoons of butter and in it brown 2 slices of onion, a chopped green pepper, and a sliced garlic clove. Stir in 2 teaspoons of salt, 2½ cups of tomatoes, ½ bay leaf, 8 peppercorns, ½ teaspoon paprika, and 1 tablespoon of brown sugar. Stir and cook for two minutes. (You can also add chopped olives or mushrooms, slivered almonds, or just about anything from the homestead garden.) Place the drained, sliced tongue in a casserole, pour the sauce over it, and bake it at 375° F. for a half-hour.

Taste this just once, and if you don't already appreciate such meats, you'll experience an awakening.

Scrapple

To make scrapple, cook head meat, heart, and trimmings until the meat can easily be removed from the bones. Grind the meat with a one-eighth-inch plate, strain the meat liquor to remove the small bones, return the ground meat to the liquor, and bring it to a boil.

Mix four parts meat, three parts liquor, and one part cornmeal (by weight). Mix the meal with some of the liquor first to avoid lumps. Then add it to the meat and the rest of the liquor. Boil it for about a half-hour, stirring to prevent sticking, and add:

- 3 oz. salt
- ½ oz. black pepper
- ¼ to ½ oz. sweet marjoram
- ¼ to ½ oz. sage
- 3 oz. ground onions
- dashes of red pepper, nutmeg, and mace

Head cheese

Head cheese is made by splitting and cleaning the head, removing eyes, ear drums, teeth, and nasal passages. Other meat can be used too, such as the heart, tail, and feet. Simmer it all until the meat can easily be removed from the bones. Dip off and retain the liquid, remove all the bones, and chop the meat fine. (It can also be ground with a coarse plate, about three-eighth inch.) Season with black and red pepper, ground cloves, coriander, sweet marjoram, and salt. Return it to the kettle, cover with the liquid that was dipped off, and boil for fifteen minutes. Then pour the mixture into shallow pans, cover with cheesecloth and weight it down. It's ready to slice and eat without further preparation.

Pigs' feet

The feet can also be made into a delicacy. They should have been cleaned when the carcass was dressed, and the toes and dewclaws removed. After cleaning and chilling, cure in a Tender-Quick pickle (by Morton Salt) made with two pounds of pickle to one gallon of water. After a week to ten days, take the feet out, wash them, and simmer them until the meat is tender. Cook them slowly. Then chill them, pack them into a tight container, and cover them with hot spiced vinegar. They can be served cold, or fried in a batter of eggs, flour, milk, and butter.

Finally, the tail

The tail can be kept by the kitchen range for lightly greasing the frying pan—merely rub it over the hot surface. When you butcher a pig, you use everything but the squeal.

Other cuts can be prepared just like you've always cooked pork, with one possible exception. The curing material used by commercial meatpackers contains a type of sweetener that helps prevent bacon from turning black when fried too quickly. Bacon that is home-cured or cured by a small slaughterhouse should be fried slowly at a low temperature in order to keep it from getting black.

Roast suckling pig

Roasting a whole pig provides the basis for a feast that can be enjoyed only by a relative few—like wealthy people, and hog-raising homesteaders. A suckling pig on a spit is a conversation piece that will be remembered by party guests for years to come as a gourmet's delight.

Preparing a forty- or fifty-pound pig is much easier than butchering a two hundred-pounder, although the process is the same. Kill it, scald it, and scrape it well. Remove the intestines, but leave the head on. Remove the eyes.

Homemade spits are easy to rig up. You could use a green sapling and a couple of forked sticks to hold it, campfire style, or a more elaborate one made of scrap metal you might have access to. There is one problem to be ready for. No matter how tightly the pig is tied to the spit, as it begins to cook it will shrink and work loose. You will find yourself turning the spit—but not the pig. The shaft of the spit needs something at right angles to it to ensure that the meat will rotate with the shaft.

You can use charcoal or, if you prefer, wood. Use a

Whole pig on a homemade spit.

cooking fire, not a bonfire; glowing coals will do a better job than crackling flames.

For more attractive results, cover the ears and tail with foil until the last hour of cooking to avoid scorching.

A meat thermometer is almost a necessity. A fifty-pound pig can appear beautifully golden and glazed and still be underdone. A pig should be roasted at 350° F. until a thermometer in the thickest part of the ham registers 170°. This can take as long as twelve hours for a fifty- to sixty-pound pig, dressed weight. While a suckling pig will cook faster, a larger hog will take even longer.

Roasting even market-sized hogs has become quite popular in recent years. In my area at least, there are several people who have built special ovens large enough to accommodate a whole hog, and they keep quite busy during the picnic season roasting pork for large gatherings.

The Hawaiian method is to dig a pit, build a fire in it, wrap the pig in banana leaves, and bury it. Having lived in Hawaii, I can attest to the succulence of such a gourmet item, but we've done as well with a spit here in the Midwest—and had roasting ears of sweet corn besides!

Chapter 12.
MEAT QUALITY

A reader of *Countryside* recently wrote, "Everybody who raises hogs on their homestead is always exclaiming how much better home-produced pork tastes, but until I raised my own, I had no idea how *much* better!" Quality ranks high on the list of reasons for raising your own pigs. However, "quality" might mean different things to different people, and it can be elusive.

For the homesteader, quality might simply mean the flavor and tenderness of the pork. Those concerned about chemicals in their food will consider the chemical-free aspect of homestead pork a qualitative factor. The commercial hog producer, meatpacker, and retailer might have different ideas.

There may be some scientific basis for the better flavor and tenderness of pork that is produced and processed right on the homestead. I became interested in this a number of years ago, when my family detected a noticeable difference between pork we processed ourselves and pork we sent to an abattoir. The pigs had the same general genetic makeup. They were fed and cared for the same, in the same environment. The only difference was that I butchered some of them myself and others were sent to a processor.

It was suggested that the processor switched pigs on me: I got back an ordinary one, and somebody else got my milk- and comfrey-fed, pampered porker. "Impossible," the processor said. Each animal is tattooed when it enters the plant, and the chances of getting a carcass to the wrong owner are about as great as sending a newborn baby home from the hospital with the wrong mother—which *does* happen once in awhile, of course. But there had to be another reason for the quality difference.

Flavor is a somewhat subjective attribute anyway. Was it possible that your involvement, the satisfaction of doing it yourself, enhanced the pork? Possibly, but for people who are very aware of this factor in home-produced tomatoes, eggs, milk and dairy products, and many other foods, the difference would have been more nigligible than the one we noticed, and of a somewhat different nature.

Then I looked into the possibility that the shipping had something to do with it. When I butcher I go out to the pen and stick the pig and he expires in a very mellow mood. In contrast, the one that gets shipped is forced onto a truck under protest, subjected to the shock of a ride at fifty-five-miles per hour, shunted off to a pen in strange surroundings, frightened by unusual sights and sounds and smells, and then driven to the killing station with an electric prod. Bad vibes all the way.

While there doesn't seem to be any data on shipping versus butchering at home—the researchers would say nobody butchers hogs at home today, and who'd pay for the study anyway?—there is some information that is probably applicable.

Donald J. Willems, of Armour and Company, works at preventing livestock losses during marketing. He suggests that hogs be sent to slaughter with a minimum of exertion to improve meat quality. Several factors are involved.

Hogs generally have continuous access to feed or are fed frequently, and their rations are concentrated. In confinement or partial confinement they don't get much exercise. Therefore they don't eat much at one time. In theory, this is supposed to improve carcass yield and quality, but in practice the effect may be just the opposite.

Willems explains that the unaccustomed exertion and excitement of shipping cause the gut content to be rapidly depleted. This in turn causes the body to draw on glycogen reserves from the muscle tissue and liver, which reduces carcass yield. "And even more important," he says, "it reduces quality. The reduction of quality is manifest by lighter color, two-tone pork and watery pork."

Confinement hogs are especially susceptible, and since more and more of the pork showing up in the supermarkets comes from confinement hogs, the difference between that pork and homestead pork is all the more noticeable. Confinement hogs generally are not afraid of humans, they're not familiar with electric prods or slappers, and therefore don't drive well. They are cautious, tend to shy away from or become curious about things other hogs ignore, and have low stamina. Having lived in temperature-controlled buildings, they are sensitive to temperature variations. And, because they are not physically conditioned for exertion, they can't take it. In spite of that they are combative, and when hogs from different finishing pens are mixed on trucks and in holding pens, fighting is common. All of this creates stress.

There's more. Some researchers feel that pale, soft, and watery (PSW) pork is related to genetic factors, and they believe the increased incidence of PSW pork in recent years is related to the emphasis on production of more lean meat and less fat. There is some evidence that PSW pork seldom appears in excessively fat, poorly muscled hogs while it appears frequently in trim, extremely meaty-type hogs. However, it doesn't appear in all meat-type hogs, which suggests that it might be possible to breed and produce meat-type hogs with a low incidence of PSW.

The direct cause of PSW appears to be a rapid and abnormally large buildup of acidity (lactic acid accumulation) within the muscle fibers of the carcass immediately after slaughter. Stress, in some as yet unknown way, speeds up the production of acidity in the meat.

According to John F. Lasley of the Department of Animal Husbandry at the University of Missouri, "PSW pork may be due to an upset in the enzyme system controlling the breakdown of glycogen (animal starch) in the muscle fibers. Certain enzymes are involved in a series of reactions which change glycogen to lactic acid and then to pyruvic acid or to certain end products depending upon the presence or absence of oxygen.

"If the enzyme which changes lactic acid does not function normally, lactic acid accumulates and causes the muscle fibers to become more acid, resulting in the PSW condition."

Research in several states, and indeed in several countries, indicates that the occurrence of PSW pork is due to heritable factors. Recent emphasis on selection for meat-type hogs may have run up against a gene responsible for PSW pork. However, in breeds where the meat-type individuals do not possess the specific gene, the PSW condition is not created.

The upshot of this is, it could probably be proved scientifically that homestead pork is better than store-bought because of better growing environment and slaughter methods. On the other hand, if you have stock that is susceptible to PSW, and especially if you ship the animal to a slaughterhouse, you run the risk of missing out on the delectable eating most homestead pork producers rave about.

Even so, this is only a part of pork quality. Much depends on the breeder—which for the homesteader means the farmer who owns the sows and selects the boars which produce the feeder pig the homesteader buys.

We mentioned earlier that hog "type" has changed in the last generation or so. When families were large and spent their days in vigorous outdoor activity, fat pork was not only tolerated: it was preferred. Not until most of us moved indoors to controlled temperatures and to lifestyles where the most vigorous activity was climbing stairs or bending down to pick the newspaper off the front porch, did housewives begin to turn to leaner cuts. (Some people might see a correlation here with the confinement hogs.)

Moreover, until the late forties, lard was an important product, and fat was worth almost as much as lean meat. Then vegetable oils replaced lard as a shortening, and synthetic detergents replaced lard-based soaps. Thereafter, hams, loins, picnics, and butts far outdistanced lard in terms of value. Since it took time and feed (money) to produce lard, hog raisers naturally endeavored to produce hogs with a greater proportion of the primal cuts, and less fat. So, for packers and the producers who sell hogs to them, "quality" today also means the proportion of lean to fat. The change was largely wrought through genetic selection, but as noted in regards to PSW, it possibly brought in other problems.

Most homesteaders probably prefer the same type of lean meat demanded by the average consumer—because they're accustomed to it, because their lives are practically as sedentary as urbanites', and because they too have been brainwashed about cholesterol. For the homesteader who wants to be self-sufficient and is interested in lard for shortening and homemade soap, even the average "lean" hog will produce enough fat for those purposes.

The homesteader, therefore, is interested in the same kind of hog the big packers look for. This means pork that is palatable: tender, juicy, and flavorful. It means attractiveness: marbeling, color, and moderate degree of fatness.

Small cuts are more popular than formerly because of smaller family size, and this has had an effect on the type and weight of animals marketed. As far as the retail market is concerned, housewives want cuts that are easy to prepare, that are tender, and that are the same on each trip to the supermarket.

Producers strive to supply this kind of pork through hog management. Because excess fat is the greatest single cause of poor quality pork, they pay special attention to breeding in order to attain meat-type animals. In addition they restrict the rate of gain to about 1.5 pounds daily after the animal reaches 125 pounds by using bulky finishing rations, and they slaughter at lighter weights than formerly, not over 220 pounds.

Measuring backfat thickness and loin eye area

In selecting breeding stock that can be expected to produce the desired meat-type carcass, the two most common management tools are measurement of backfat thickness and loin eye area.

Backfat thickness at two hundred pounds should not exceed 1.4 inches in gilts and 1.25 inches in boars. With practice, backfat can be measured visually with some degree of accuracy by taking into account flabbiness of jowls and bases of hams, countersunk tails, broad and shelflike backs, excessively wrinkled shoulders, and lack of trimness and firmness in underlines. A much more accurate assessment can be made with an ultrasonic device that actually measures the fat layer with a process much like sonar.

There is also a backfat probe, a tool whose function can be duplicated on the homestead with a knife and a small metal ruler. Probing backfat causes no permanent injury and not much discomfort.

The knife blade is wrapped with tape ⅜ inch from the tip to keep it from penetrating too deeply. It is jabbed sharply through the skin at right angles to the body, the ruler is inserted and forced through the fat to the muscle, and the clip on the ruler which is made for this purpose is forced down to the skin. Probes are made in three locations: behind the shoulder directly above the front elbow but back about an inch; at the last rib; and halfway between the last rib and the base of the tail. All three are made 1½ to 2 inches to either side of the center of the back.

Hogs should be probed when they weigh from 175 to 225 pounds, although measurements are always based on a 200-pound live weight. In order to adjust the measurement to 200 pounds live weight, a conversion factor must be used.

Live weight	*Factor*	*Live weight*	*Factor*
175	1.070	205	.987
180	1.056	210	.974
185	1.043	215	.961
190	1.028	220	.959
195	1.014	225	.953
200	1.000		

To obtain the average backfat thickness, total the three probes and multiply the result by the above factor corresponding to the weight of the hog.

If the feeding program is based on 180 days (six months), it is also necessary to adjust weights to the same age in order to make an accurate comparison. This can be accomplished by adding 2 pounds to the hog's actual weight for each day of age less than 180. If the pig is more than 180 days old, subtract 2 pounds for each day over 180. For adjusting weights to 154 days or five months, 1.65 pounds is used instead of 2 pounds.

USDA grades are based on backfat thickness correlated with carcass length. U.S. No. 1's can range from 1.3 to 1.6 inches of backfat depending on carcass length and live weight. U.S. No. 2 carcasses have a maximum average backfat thickness increasing from 1.6 to 1.9 inches with increases in carcass length from 27 to 36 inches. For U.S. No. 3, maximum backfat increases from 1.9 to 2.2 inches with carcass length increases from 27 to 36 inches, and U.S. No. 4 has a lower yield of lean cuts than U.S. No. 3.

A fifth grade, U.S. Utility, indicates carcasses that have a lesser development of lean than is required for the four top grades. In addition, any carcass that is soft and oily or does not have acceptable belly thickness is graded U.S. Utility regardless of any other quality characteristics.

Based on carcass length and backfat thickness, the four top grades are expected to yield the following proportion of the four lean cuts:

Grade	*Expected yield of ham, loin, picnic shoulder, and butt, based on chilled carcass weight*
U.S. No. 1	53% and over
U.S. No. 2	50 to 52.9%
U.S. No. 3	47 to 49.9%
U.S. No. 4	Less than 47%

Loin eye area is another measure of quality. Generally a tracing is made of the loin eye area, and a special plastic grid is placed over the tracing. The grid has ten squares per inch. The number of squares within the loin eye are counted, and that total divided by ten gives the number of square inches within the loin eye.

While backfat reflects carcass leanness, loin eye area is the single most accurate indicator of total carcass muscling. The national certification standard is 4.50 square inches. (It's interesting to note in passing that at the 1973 National Barrow Show, 45.8 percent of the animals were disqualified because they failed to meet the 4.50 standard.)

Any homesteader can produce good pork accidentally, but those who strive to do so, who know what quality pork is and work at producing it, will have better products more consistently.

W.L.C.Co.
GUARANTEED

POSTSCRIPT

We have discussed—and not entirely jokingly—keeping a pig on your patio. And just today as I finished the last chapter, I ran across a newspaper clipping that indicates the concept might not be as amusing or farfetched as it may seem.

The headline on the clipping is "Italians Raise Mini-Pigs on the Balcony." The item claims that because of Italy's increasing meat shortage, to say nothing of high prices, people are starting to raise their own pigs. There is a miniature strain (used for laboratory work in this country) that is only eighteen inches long and a foot high. Piglets of this strain are selling for $50 each, according to the article. They are fed kitchen scraps—and are kept on balconies in the cities. In a matter of weeks the mini-pig is fully grown, and supplies a family with meat for the better part of a week.

Now then, about that patio . . .

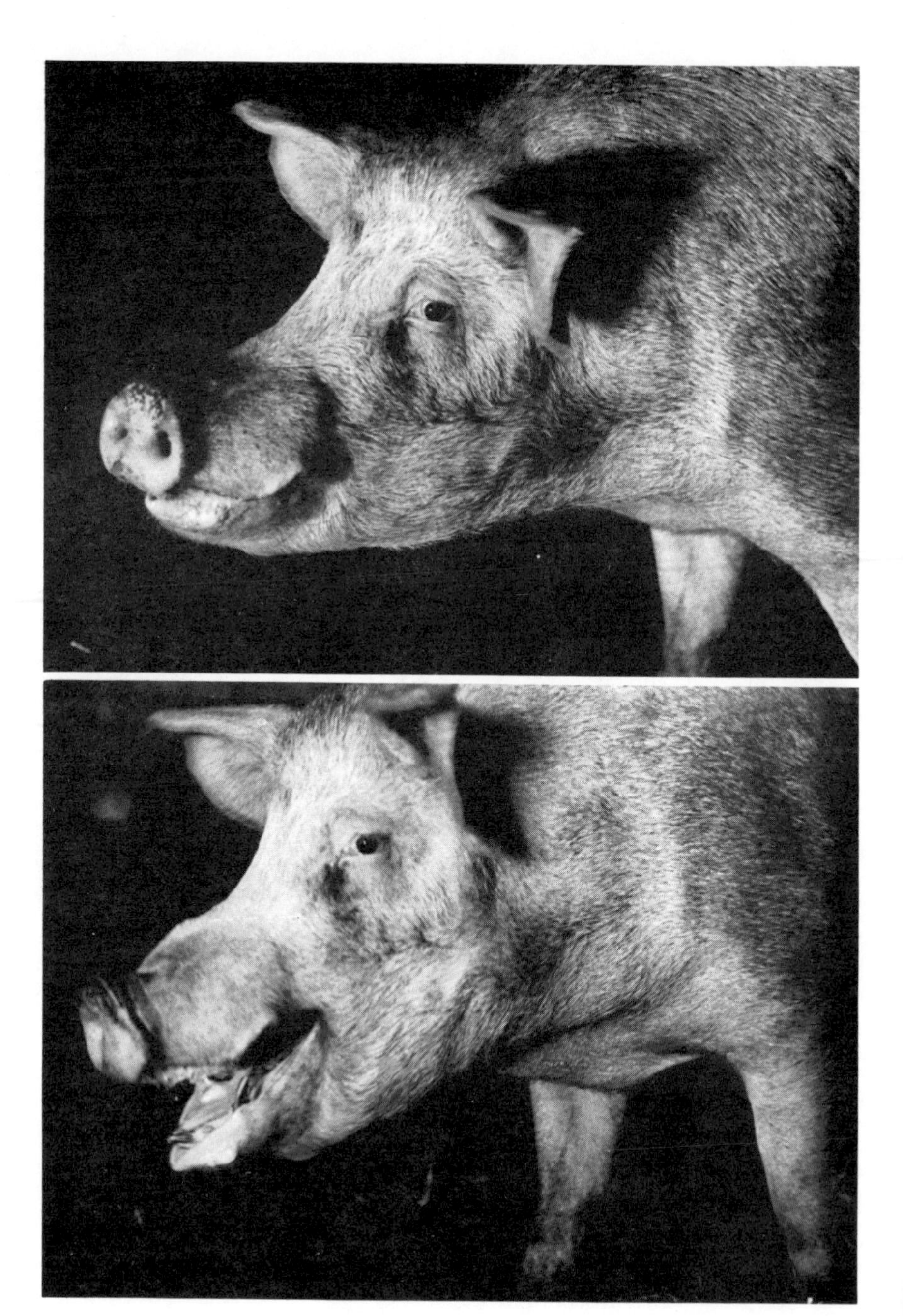

APPENDIX
A. FARROWING

Strictly speaking, this book is not designed for anyone interested in farrowing sows. We can speak at least halfway seriously about keeping a pig on your patio, but a sow by the swimming pool is a bit much.

We've been discussing producing pork primarily for home consumption. A sow doesn't fit this concept very well. Any sow worth keeping should wean at least seven pigs at least twice a year, which means at least fourteen pigs or 1½ tons of pork per year.

Anyone who is going to raise hogs commercially, on the other hand, will want to study the books and manuals written expressly for commercial producers, and probably gain as much on-the-job training as possible. Nobody in their right mind takes up hog raising without a great deal of planning and preparation.

However, once again there is a fuzzy gray area in between—an area many people are finding themselves in. They are neither fish nor fowl, neither self-sufficient homesteaders nor commercial farmers in the usual sense of the term.

(It should be noted here that the common idea of a modern farm being huge, automated, and probably incorporated, and the small family farm being nothing more than a nostalgic figment of the past, isn't quite true. While the vast bulk of our food and fiber is produced by only a handful of large operations, the number—and importance—of the smaller units is still formidable, in spite of the efforts of the USDA to wish them away.)

No two people are in exactly the same circumstances, and there can be many good reasons why someone who starts out to raise a family pig ends up with a couple of sows. One reason is probably the added sense of independence; the *entire* pig is produced on the homestead—a pleasing notion to some, but pretty poor economy. Other reasons might be the desire for additional income or the availability of cheap sows. (Many commercial breeders only keep sows for a few farrowings. The price per pound for sows is less than for butcher hogs, but they don't "depreciate" any more with age.) A farrowing operation takes less space and feed than a feeder operation, assuming of course that the babies will be sold shortly after weaning.

The list of disadvantages is much longer. A good deal more knowledge and management is required for sows than for feeders. A farrowing operation does not fit in nearly so neatly as a part-time job as does feeding out hogs; many a farmer sits up all night with a farrowing sow, but he doesn't have to drive to town in the morning and face a boss who could care less about the pork stork. And pigs farrow during the day, too, when the part-timer won't be around.

The investment in sows is greater than the investment in an equal number of feeder pigs, and more equipment is needed. These two factors combined with the greater skill and knowledge required mean that the risk is greater. A homesteading friend of mine keeps just a few sows. One farrowed last week and lost her entire litter. That means the man lost about four months' work and feed; a sow eats from four to six pounds of feed a day.

Then there is the added expense and trouble of keeping a boar: original cost, housing, feed, and care. If you're feeding one sow and one boar, the initial cost of a litter is effectively doubled. (Young boars can be used for about a dozen services per breeding season; those two years old or older can serve forty to fifty females.)

It has been estimated that as many as 20 percent of all gilts never get bred, which, again, raises the costs of the farrowing operation. Add to this the normal mortality of baby pigs (some people say 25 percent of the pigs born never reach market), the additional diseases that affect sows and suckling pigs, and similar considerations, and the sow enterprise begins to look much less attractive to people who thought it seemed like an easy way to make some cash.

Having said all this, I'll concede that some people do very well farrowing sows on a small scale. But I do think most of them have the experience that isn't found in books. Of course, we all have to gain experience *somewhere,* and if you didn't learn working with your dad's pigs and want to keep sows anyway, you'll just have to learn on your own. Feeding out a couple of hogs will provide at least a basic familiarity with the animals, and a dedicated person *can* learn from experience.

The normal procedure is to remove gilts from the feedlot at two hundred pounds. Those gilts saved for breeding are selected long before, preliminary selection sometimes even being made at birth. All the heritable characteristics are taken into consideration (see the chapters on breeds) as well as the records of the gilt and her brothers and sisters. She obviously must be sound and healthy.

Gilts are bred to farrow at one year of age. (The gestation period for swine is about 114 days or, as an easier way of remembering it, three months, three weeks, and three days.) They are "flushed" ten days to two weeks before breeding. This is the practice of increasing their feed to stimulate ovulation. Rations that were reduced upon removal from the feedlot to prevent excessive weight gain and fatness are increased to six to seven pounds per day.

Swine come into heat every eighteen to twenty-four days and remain in estrus two to three days. It's often said that a swollen and reddish vulva is a sign of heat, but this isn't always reliable. Other even less noticeable signs (in some sows) are restlessness, mounting of other sows, and loud grunting that sounds almost like barking.

Gilts should pass over two to three heat periods before being bred for the first time. Hand mating (introducing a gilt or sow in heat to a serviceable boar) is preferable to pen breeding, where a boar is run with several females.

A high level of feeding immediately after breeding is said to be detrimental to the implantation of embryos in the uterine horn, so sows and gilts should be allowed only limited feed after breeding. Four pounds of a complete ration is sufficient.

The most critical period in gestation is between the eighteenth and thirty-second days. If a pregnant sow gets sick during that time she is likely to have a small litter with weak pigs. A radical change in feeding or environment during that period may also result in poor litters.

There are several common methods of limiting feed intake of sows. Feeding only four pounds per head might work in some herds, but more than likely one sow will be able to eat four pounds in fifteen minutes while another will take thirty or forty minutes. One will be overfed, the other underfed. This problem can be overcome by stall feeding. Feeding stalls for sows are twenty inches wide and about eight feet long. When a sow goes in to eat, she can't be chased away by a more gregarious neighbor.

Other breeders prefer to provide full feed, but a special, low energy ration. The sow is allowed to eat as much as she wants, but because of the formulation of the ration, she does not gain more weight than is desirable. Such rations may include alfalfa, a larger proportion of high fiber grains such as oats, and even silage. (Sows and gilts will eat 10 to 12 pounds of silage a day, along with 1½ pounds of a balanced supplement containing protein, minerals, and vitamins.) Ground whole ear corn is suitable for sows in gestation, but rations containing cobs are not high enough in energy for growing pigs or lactating sows.

The only reliable pregnancy test is the farrowing of a litter. There are several very expensive machines on the market that are said to do the job, but they require rather large herds to justify. Sows that miss but still show up at the feed trough are definitely a drain on the hog enterprise. On occasion female swine show every external symptom of pregnancy until two or three weeks before their scheduled delivery date, when they lose the signs and do not farrow. Nobody has an explanation.

Most sows are routinely wormed during the last month of gestation, treated for lice amd mange, and in some cases given injections and potions of various kinds depending upon the breeder's experience with diseases and problems. Several days before the due date, the sow is scrubbed with soap and warm water and moved to a disinfected farrowing pen. The pen can be disinfected by using one pound of lye to fifteen gallons of scalding hot water.

Farrowing pens can take various forms. In large and modern commercial piggeries the farrowing crate is standard equipment. This is basically a cage of sturdy metal construction which allows a sow to lie down or stand up but little more. Its purpose is to prevent the sow from lying on her babies—the number-one cause of baby pig losses. The first week is the most critical in this regard, and some producers put the sow and litter in a conventional pen after that.

A regular pen can be used for farrowing, but a guardrail should be constructed around the perimeter to prevent "overlaying" (what the professionals call it when a sow squashes her babies). A guardrail can be as simple as a 2-

by-6 attached to the pen walls so it's 8½ inches from the floor and so it extends 8 inches from the wall itself. The rail obviously has to be sturdy enough to withstand the pressure of a sow, but since one-third of baby pig deaths before weaning are due to overlaying, some form of protection is well worth the effort.

Feed should be as laxative as possible from the time the sow goes into the crate (or equivalent) until the pigs are several weeks old. This is said to help milk flow by preventing MMA (mastitis-mertritis-agalactia). Drastically reducing the feed can cause constipation, which can trigger a lack of milk flow. When a sow goes off milk, try feeding her raw liver.

The amount of milk a sow produces depends on many factors including breed, age, litter size, number of previous litters, feed, frequency of nursing, and others. But as an average, a sow produces eight to twelve pounds of milk per day soon after farrowing, reaching a peak of fourteen to twenty pounds a day about a month later. (Eight pounds of milk equals a gallon.)

Other causes of problems with milk flow can be extreme heat or cold, diarrhea or constipation, sudden environmental changes, systemic diseases, hormonal imbalances, and rations or feeding practices. According to some authorities, most cases of MMA now seen seem to involve bacterial infection, and these persons speculate that modern intensification of swine production has something to do with it.

Attend the sow at farrowing and give assistance if needed. Generally the main job is to clean up, dry off, and warm up the newborn pigs as they arrive. Stillborn pigs can often be revived by mouth-to-mouth resuscitation. Place a plastic funnel over the pig's nose and mouth, and blow into the small end. Or rub the pig briskly on both sides of the rib cage.

There is no "normal" farrowing time. In an English study it was found that births usually occur within 15 to 60 minutes after the sow lies down after building her nest.

After the birth of the last pig, she usually stands up to urinate, and this is when most baby pigs are crushed. Average parturition time in this study was 2 hours and 53 minutes, but it ranged from 25 minutes to 8 hours and 55 minutes. Expulsion of the cleanings (afterbirth) occurred anywhere from 21 minutes to 12½ hours after the birth of the last pig, with an average of 4 hours.

If navel infection is a problem, follow the correct procedure: tie off the navel about 1½ inches from the belly immediately after birth; then cut the cord and paint the stump with iodine or a mixture of iodine and glycerine.

Remove the afterbirth when it is expelled. Some breeders feel that allowing the sow to eat the afterbirth (which is her natural instinct) sometimes triggers cannibalism.

Even so, the occasional mother will disown, or even destroy, her young. The best treatment for this type of sow is to remove the babies as soon as they are born, keeping them in a warm box. They can go three to four hours without nursing, if necessary. Quiet the sow by scratching her behind the ears, rubbing her udder, and talking to her. When she is lying down, let a couple of the pigs nurse. If the sow remains quiet, let them all nurse. Then remove them again. If this is done a couple of times, the sow will probably accept her litter.

Newborn pigs require temperatures between 80° and 90° F. This is generally supplied in a hover, or triangular brooder secured in one corner of the pen. This brooder need be nothing more elaborate than a used door nailed to two walls about nine inches off the floor. Never suspend a heat lamp by its cord, but by a separate chain. Even then, have some sort of protective shielding in case it should fall to the bedding.

In mild temperatures soft deep bedding will be sufficient without supplemental heat, as the babies will burrow in, huddle together, and keep quite warm from their body heat. But in any event, bedding for baby pigs should always be clean, dry, and away from drafts.

Watch the pigs to make sure the runts are getting milk.

Oftentimes the weaker members of the litter are shunted aside by the stronger ones to the point of starvation. Do not remove them from the sow, but provide them with enough energy to keep them going until a sucking order has been established. This is accomplished with tube feeding. One method uses a 20 cc. hypodermic syringe, about a foot of soft rubber tubing with an inside diameter small enough to fit snugly over the end of the syringe, and a clamp (or a helper) to hold the syringe while the pig is being fed.

Mix 1 quart of cow or goat milk, ½ pint of half-and-half, 4 tablespoons of white Karo syrup, 1 egg, and if you wish an antibiotic or baby vitamins.

Hold the pig by the head (grasping the back of the neck), carefully guide the tube through the mouth into the esophagus, and force the milk into the stomach. Small pigs should get 15 to 20 cc. every four hours until they are strong enough to compete with their brothers and sisters.

Gradually increase the sow's feed after farrowing, allowing about two weeks to get back to full feed.

Baby pigs deplete their store of iron by ten to fourteen days of age. Therefore, if sows are on concrete, pigs should have iron shots between their third and fifth day to anticipate this shortage. Should you wait longer than five days before replenishing the iron supply, expect to see signs of anemia by about two weeks of age.

Other management tasks at this point might include needle teeth clipping and tail amputating if you intend to sell feeder pigs. Pigs with docked tails are commonly bringing higher prices when sold to farmers who feed out hogs in confinement. Needle teeth are very sharp and can cause pain and discomfort to the sow. They are of no benefit to the pig. Cut them with a small wire cutter or with forceps designed for the job. Only the tips are removed, to avoid injury to the gums or jaws. There are eight of these needle, or black, teeth: two on both the upper and lower jaws on both sides of the mouth.

Ear notching (to identify pigs) is done at the same time needle teeth are clipped. Purebred herds need this identification for registration, and commercial herds need identified pigs kept for breeding stock.

The sow and litter can be moved to clean pasture when the pigs are about two weeks old, by which time they will of course have been castrated. Creep feeding is usually started then, too. This is the practice of providing a special pelleted ration to the babies in enclosures too small to admit sows. Early gains are the cheapest; pigs that can put on weight effectively at this age will do better, and produce cheaper pork.

Weaning

Pigs are weaned anywhere from a few days of age to two months. Faster weaning naturally requires more experience and management, as well as special feeds. The biggest advantage is to get the sow back into production, a prime goal in commercial herds. The organic-minded farmer probably isn't in so great a rush.

Weaning is most commonly accomplished at five to eight weeks of age. If creep feed has been available, very little if any setback should be noticeable. Decrease the sow's feed a few days before weaning, and increase the bulk of the ration until the udder has dried up. Complete and final separation is advisable, meaning that the sow and pigs shouldn't even be able to hear or see each other.

Most sows will come in heat three to five days after farrowing, but they won't conceive if bred at that time. Very early weaning (at less than a week of age) permits the sow to be rebred at thirty to forty days after farrowing. Sows will not come into heat during lactation except for that first "false" heat, so weaning a month or two later will result in the first heat being delayed a month or two. The importance of this delay depends upon your goals and management methods.

When the sow is bred again, the cycle starts all over. In commercial herds not many sows are kept past their second to fourth litters. By then most of them are large enough and strong-willed enough to be troublesome when run with younger and smaller sows or gilts, and they are rougher on equipment. Older sows can also be expected to have foot and leg troubles, and later, problems with seeing, hearing, and teeth.

B. CASTRATING

The reasons for castrating are to prevent uncontrolled breeding, and to prevent the development of boar odor and flavor in the meat. Because of the latter, boars are heavily penalized in the market.

Castrating pigs is like many other skills involved in animal husbandry in that there is little general agreement on the best procedure, and the best way to learn is by working with an old pro.

Some people claim the best time to castrate pigs is from four to seven days of age. Others point out that the testicles are small then, which makes the job harder, and there may be a layer of fat in the scrotum which will not be there later, which also makes the operation more difficult. However, pigs at that age are as clean as they'll ever be; they are probably in better condition than they'll ever be; they are getting excellent feed (their mother's milk); and they will suffer less shock.

Castrating older pigs is easier for many people. However, castrating pigs five to seven weeks old will keep them from gaining weight for as much as a week afterwards; death losses of 2 percent and more are not uncommon; and the pigs are harder to catch and hold. Older pigs will suffer even greater setbacks and greater losses.

Castration can be performed with a good sharp jackknife, a razor blade, or similar cutting tool. Other items needed are a bottle of equal parts of iodine and glycerine and a supply of cotton swabs. Keep all equipment sterile and on an instrument pan.

Here is one common way to proceed. Have an assistant grasp the pig by its back legs and place it on the work surface. A table or bench will do, although people who castrate a great many pigs use a castrating rack, which is simply a V-shaped trough. The assistant pulls the back legs forward along the stomach and at the same time grasps the front legs.

Swab the scrotal area with the iodine and glycerine. Make an incision vertically (lengthwise to the pig) over the center of the first testicle. This is more easily accomplished if the testicle is forced outward against its skin between the thumb and forefinger. The incision should be made with one clean stroke, all the way through the skin layers to the testicle itself. The testicle will then push out from the scrotum.

Follow the same procedure on the other testicle. When both testicles are exposed they can be pulled free with only the spermatic cord holding them. Or, stretch out this cord and cut it as far up as possible.

The scrotum should then be treated with a clean swab containing iodine and glycerine, or sprayed with a disinfectant. Place the pig in a clean pen away from uncastrated pigs.

There are, as mentioned, other methods. Some people have the assistant hold the pig by the hind legs and perform the operation in that position. Others place the pig on the ground, hold the front legs and head down with one knee, and work without an assistant. Some people make only one incision, between the two testicles. No one way is "best" unless it's best for you.

C. THE LEGALITIES

From a technical standpoint, almost every family in America that doesn't live in a high-rise apartment building could raise its own pork. All that's required is about eighty square feet of space. There doesn't have to be a problem with odor, rodents, noise, or any other nuisance.

Most Americans don't want to raise their own pork, of course, but there's no real reason it couldn't be done in the average yard, even in our cities. But alas, many people who do want to raise their own pork, and who have considerably more space than the average city lot, can't. The law won't let them.

Zoning regulations are a hobgoblin for many homesteaders. Most municipalities have restrictions against even chickens and rabbits, and some homesteaders are alarmed to learn that such restrictions exist even in what they consider rural areas. Some thoughtful people believe that attitudes resulting in such laws are going to have to change in the years ahead as food becomes more scarce and expensive, and in some cases restrictions have already been relaxed as increasing numbers of people who want to produce their own feed have challenged them.

If misinformed health authorities and those misguided souls who think food is produced in the back room of the A&P get upset about innocuous rabbits, imagine how they react to hogs!

Swine have always had a bad press, starting with the Old Testament. Our language is full of references to swine, almost all of them bad. "Bringing home the bacon" is one favorable expression that comes to mind, but it's grossly overshadowed by "filthy as a pigpen," "filthy swine," and just the implications alone in the epithets "pig," "swine," and "hog." Little boys with poor table manners have always been told not to "eat like a pig." Gluttonous people "make hogs of themselves." A teenager's unkempt bedroom "looks like a pigsty."

A Pennsylvania judge wrote that, "It is recognized by all authorities . . . that the keeping and raising of hogs . . . is likely to be offensive, and has a tendency to endanger the health and destroy the comfort of residents in that vicinity." That must mean that I'm no authority because I don't recognize that at all. Pigs smell, but some of us don't find it offensive. Who's to say what olfactory reactions are induced by conditioning? Don't we all know children who think flowers "stink" but who grow up thinking differently because flowers are *supposed* to smell good? Remember the story of the World War II GI in France who thought the Germans had developed a new, foul-smelling poison gas, when actually it was the delightful aroma of new-mown hay?

Old ingrained ideas don't die easily though, and it's unlikely that most people will change their minds about pigs. This has very serious implications for the homesteader.

Zoning laws vary tremendously from place to place, and the only way you'll find out how they affect your homestead is to personally check into local regulations. Even then your course of action might not be clear.

A New York court once closed down a piggery because, "Rats are living on the property in numbers, and they would tend to stray onto the public highway and onto adjoining lands." That hog producer isn't likely to get much sympathy from anybody. In a recent case in the state of Washington, a family was forced to get rid of not only their pigs, but goats and other livestock as well. The judge described the defendant's yard as, "A malodorous pig-sty, bounded by sagging, decrepit, patched-up fencing festooned with garbage, enclosing hogs wallowing in muck and mud—a pen littered with garbage, refuse, debris, and filth against a backdrop of junked automobile bodies occupied by numerous dogs as sleeping quarters—a place from which . . . a stench would naturally be deemed to emanate." In this case *(State* v. *Primeau)*, the defendant had been raising animals for ten years before a zoning law reclassified his neighborhood, and under ordinary circumstances a new ordinance would not have interfered with activities that took place before the ordinance was passed. But in this case too, who is going to have much sympathy for people who raise animals like that?

On the other hand, a Missouri court in 1972 came to the opposite conclusion. This case involved twenty-nine hogs being kept within the city limits. The nearest neighbor testified that she had never been bothered by dust, debris, odor, or noise, and evidence showed that the defendant's premises were "reasonably clean." The pigs won.

These cases illustrate the urgent necessity for designing hog facilities that are neat and attractive, and keeping them that way. In an urban setting this requirement will even exceed the usual sanitation goals of one who wants to raise healthy hogs for home use. The demands of aesthetics are often harsher than the demands of practicality.

But even this is not always the answer, and the law is not always this clear-cut. For example, in some places pigs are not even allowed in agricultural zones! A court in Massachusetts ruled (in 1943) that "the maintenance of a piggery is different from ordinary farming," and that therefore zoning boards have the right to place restrictions on such operations, or prohibit them altogether. A court in New York made a similar observation in 1954. While agreeing that farmers may keep "a reasonable number" of livestock—whatever *that* is in this age of agribusiness and bovine and porcine concentration camps—the court also felt that, "The extensive raising of pigs to be sold and disposed of shortly after birth is not among the usual accessories to a farm."

While activities that preexist a new regulation might be permitted to continue, in some cases those regulations might still cramp your style. There have been several cases where preexisting conditions were allowed to continue, but when the parties operating under a grandfather clause expanded, or even just improved their facilities, the courts blew the whistle. No doubt this could also be expected to affect a homesteader who raises goats and poultry under a grandfather clause, and wants to expand into hogs.

Some jurisdictions permit swine, but only in certain limited numbers. While these rules shouldn't bother homesteaders, their effect is obvious on farmers or anyone who contemplates increasing hog numbers. Some of these laws seem harsh, and even silly. (How many homesteaders would think of checking to see if hogs are allowed in an agricultural zone!) There are several implications for homesteaders in all of this.

First, if you don't want to tangle with the law, be aware of local zoning regulations.

Second, if you do raise hogs, keep them neat and tidy! If you don't, you may jeopardize your own hog production as well as that of others in your zoning area—and perhaps even husbanders in far-distant places when lawyers in similar litigation use you and your court case as a bad example.

And third, there is a real need to actively work for sensible zoning legislation in this country. Homesteaders, especially, shouldn't let themselves be railroaded by supersophisticates who seem to think that they'll always be able to buy food at the supermarket—even after no one is allowed to produce it anymore.

Food is *not* produced in potato chip factories, or even in meatpacking plants. Homesteaders are well aware of this. Can there be a right more basic than the right to produce your own food? Isn't that a right worth working for?

D. THE NITRATE/ NITRITE ISSUE

Many people who are concerned about the quality of food they eat are uneasy about sausage and cured meats that contain nitrite or nitrate. Yet virtually all curing and sausage recipes call for saltpeter, which is potassium nitrate. These people ask, "Is saltpeter really necessary? What does it do? What happens if I don't use it?" That these questions are not easily answered is amply demonstrated by the confusion surrounding these substances as this is written in early 1976.

For the homesteader to make an intelligent decision, we should go back to the beginning and examine how the curing process takes place. Meat has been preserved with salt for ages, of course. At certain levels, salt prevents the growth of some types of bacteria that are responsible for meat spoilage. Salt also has a drying effect on meat, and most bacteria require substantial amounts of moisture to live and grow.

It was discovered that salt from certain areas imparted a special flavor and color to cured meat, but it wasn't until the turn of the century that nitrate, which is present in some salt, was isolated as the factor responsible. Still later it was determined that nitrate is changed to nitrite by bacterial action during curing and storage and that the nitrate itself had no curing effect.

In sum, nitrite in meat insures against the development of botulinal toxin (botulism), it develops the cured meat color and flavor, it retards the development of rancidity during storage, and it inhibits the development of a warmed-over flavor.

Ordinary salt, of course, also has an effect on certain types of bacteria and on flavor. Sugar is added to reduce the harshness of the salt. Spices and smoking are for flavor, although smoking can also be a preservative.

Nitrites and nitrates are toxic, just as salt is. The fatal dose of potassium nitrate for adults is in the range of 30 to 35 grams, and of sodium nitrite, 22 to 33 milligrams. Yet to get 22 milligrams in a single dose, a 154-pound adult would have to eat 18.57 pounds of cured meat containing 200 ppm sodium nitrite. Actually that figure should be tripled, because nitrite is rapidly converted to nitric oxide during the curing process. A person who eats that much at one sitting is obviously going to have problems that will overshadow any nitrite or nitrate toxicity: if nothing else, he'd die of salt poisoning first.

The 200-ppm level has been the maximum permitted by the USDA as of early 1976. This amounts to one pound of sodium nitrite to five-thousand pounds of meat or, in the cure, two pounds per hundred gallons of pickle brine or one ounce per hundred pounds of meat using a dry cure. (These figures are lower than those given in old-fashioned recipes such as those used in this book.)

But even if toxicity is not considered a major problem, some people will consider that compounds known as nitrosamines are. Under certain conditions not yet fully understood, the natural breakdown products of proteins known as amines can combine with nitrites to form nitrosamines. There are many different types of nitrosamines, some of them carcinogens. And that's where the concern and confusion enter.

The actual levels of nitrosamines in cured meat products are not known, and many variables influence these levels, such as the concentration of amines in the meat, amount of nitrate added, type and amount of other ingredients, actual processing conditions, and length and temperature of storage. It is unknown at what levels, if any, nitrosamines are formed in humans after they eat cured meat products. And no one knows what constitutes a dangerous level in meat or in humans.

While nitrosamines are considered a definite potential hazard to human health, there is no scientific or medical evidence to indicate that meat produced and cured commercially in this country contains or produces nitrosamines at levels that are carcinogenic or mutagenic.

Two asides are in order. First, it has been found in laboratory tests that heating cured meat products affects nitrosamines. One study examined the effect of frying conditions on nitrosopyrrolidine (one of the nitrosamines) formation in bacon: when bacon was fried at 210° for ten minutes (raw), at 275° for ten minutes (very light), or at 275° for thirty minutes (medium well), no conclusive evidence of nitrosopyrrolidine could be found; but when bacon was fried at 350° for six minutes (medium well), 400° for four minutes (medium well), or 400° for ten minutes (burned), nitrosopyrrolidine formation was measured at 10, 17, and 19 parts per billion.

Secondly, it may or may not be significant, but there is little doubt that most Americans are exposed to nitrites and nitrates from sources other than cured meats. All plants contain nitrates, for example, and toxic amounts may be present when the soil is rich in nitrogen and normal growth is hampered by abnormal conditions. Water might be of even greater significance: according to a study mentioned previously, 40 percent of the wells tested in central Missouri had excessive amounts of nitrate.

What does all this mean for the homestead hog raiser?

If saltpeter—potassium nitrate—is used for its preservative properties, the choice lies between the very minimal risk of cancer and the much greater risk of botulism. However, if you cure and smoke your meat and then freeze it (as opposed to hanging it in the springhouse or similar location as the old-time homesteaders did), botulism is less of a threat. Hams and bacons handled this way are not aged, of course, but saltpeter isn't required.

Instead of having the packing industry and government making the decision, each homesteader will have to face it for himself. But then, isn't that what homesteading is all about anyway?

INDEX